Simplified Design of
Vol Converters

Con verters

Simplified Design of Voltage-Frequency Converters

John D. Lenk

Newnes

Boston Oxford Johannesburg Melbourne New Delhi Singapore

Newnes is an imprint of Butterworth–Heinemann.

Copyright © 1997 by Butterworth–Heinemann

 A member of the Reed Elsevier group

 Recognizing the importance of preserving what has been written, Butterworth–Heinemann prints its books on acid-free paper whenever possible.

 Butterworth–Heinemann supports the efforts of American Forests and the Global ReLeaf program in its campaign for the betterment of trees, forests, and our environment.

Library of Congress Cataloging-in-Publication Data
Simplified design of voltage-frequency converters / John D. Lenk
   p.   cm.
Includes index.
ISBN 0-7506-9654-0 (alk. paper)
1. Voltage-frequency converters—Design and construction.
I. Title.
TK7872.V54L46   1997
621.3815'322—dc21   97-15943
                                                          CIP

British Library Cataloguing-in-Publication Data
A catalogue record for this book is available from the British Library.

The publisher offers special discounts on bulk orders of this book.
For information, please contact:
Manager of Special Sales
Butterworth–Heinemann
225 Wildwood Avenue
Woburn, MA 01801–2041
Tel: 617-928-2500
Fax: 617-928-2620

For information on all Newnes electronics publications available, contact our World Wide Web home page at: http://www.bh.com/newnes

10 9 8 7 6 5 4 3 2 1

Printed in the United States of America

Greetings from the Villa Buttercup!

To my wonderful wife, Irene: Thank you for being by my side all these years!
To my lovely family: Karen, Tom, Brandon, Justin, and Michael.
And to our Lambie and Suzzie: Be happy wherever you are!
And to my special readers: May good fortune find your doorway,
bringing good health and happy things. Thank you for buying my books!

To Karen Speerstra, Jo Gilmore, Duncan Enright, Joan Dargan, Philip Shaw,
Pam Boiros, Elizabeth McCarthy, the Newnes people, the UK people, and the
EDN people: A special thanks for making me an international best seller, again
(this is book 88).

Abundance!

Contents

Preface

This book has something for everyone involved in electronics. Regardless of your skill level, this book shows you how to design and experiment with voltage-frequency converters, both voltage-to-frequency and frequency-to-voltage.

For experimenters, students, and serious hobbyists, the book provides sufficient information to design and build voltage-frequency converter circuits "from scratch." The design approach here is the same one used in all the author's best-selling books on simplified and practical design.

The first chapter provides the basics for all phases of practical design, *including test and troubleshooting of completed circuits.* The remaining five chapters include worked-out design examples that can be put to immediate use.

Throughout the book, design problems start with guidelines for selecting all components on a trial-value basis. Then, using the guideline values in experimental circuits, the desired results (compliance, full-scale frequency, linearity, leakage, reference voltage, scale factor, and so forth) are produced by varying the experimental component values, if needed.

If you are a working engineer responsible for designing voltage-frequency converter circuits, or selecting IC converters, the variety of circuit configurations described here should generally simplify your task. Not only does the book describe converter-circuit designs, but it also covers the most popular forms of voltage-frequency converter ICs available. Throughout the book, you will find a wealth of information on voltage-frequency ICs and related components.

Chapter 1 is devoted to basic voltage-frequency converters, particularly those circuits found in IC form. This information is included for those who are not completely familiar with voltage-frequency conversion, and for those who need a quick refresher. The descriptions here form the basis for understanding operation of the many ICs covered in the remaining chapters. Such an understanding is essential for simplified, practical design. The chapter concludes with test and troubleshooting information for all types of voltage-frequency converters.

Chapter 2 covers simplified-design approaches for Raytheon voltage-frequency converters. All of the general design information in Chapter 1 applies to the examples in this chapter. However, each voltage-frequency IC has special design requirements, all of which are discussed. The circuits in this chapter can be used immediately the way they are, or by altering component values, as a basis for simplified design of similar voltage-frequency conversion applications.

Chapter 3 provides coverage similar to that found in Chapter 2 for Analog Devices voltage-frequency converters.

Chapter 4 provides coverage similar to that found in Chapter 2 for EXAR voltage-frequency converters.

Chapter 5 provides coverage similar to that found in Chapter 2 for National Semiconductor voltage-frequency converters.

Chapter 6 provides coverage similar to that found in Chapter 2 for miscellaneous voltage-frequency conversion circuits, as well as for current-to-frequency conversion applications.

Acknowledgments

Many professionals have contributed to this book. I gratefully acknowledge their tremendous effort in making this work so comprehensive—it is an impossible job for one person. I thank all who contributed, directly or indirectly.

I give special thanks to Alan Haun of Analog Devices, Syd Coppersmith of Dallas Semiconductor, Rosie Hinejosa of EXAR Corporation, Jeff Salter of GEC Plessey, Linda daCosta and John Allen of Harris Semiconductor, Ron Denchfield of Linear Technology, David Fullagar and William Levin of Maxim Integrated Products, Fred Swymer of Microsemi Corporation, Linda Capcara of Motorola, Andrew Jenkins and Shantha Natrajan of National Semiconductor, Antinio Ortiz of Optical Electronics, Lawrence Fogel of Philips Semiconductors, John Marlow of Raytheon Company Semiconductor Division, Anthony Armstrong of Semtech Corporation, Ed Oxner and Robert Decker of Siliconix, Amy Sullivan of Texas Instruments, and Diane Freed Publishing Services.

I also thank Joseph A. Labok of Los Angeles Valley College for help and encouragement throughout the years.

Very special thanks to Karen Speerstra, Jo Gilmore, Duncan Enright, Joan Dargan, Philip Shaw, Elizabeth McCarthy, Drew Bourn, Pam Boiros, Karen Burdick, Dawn Doucette, Laurie Hamilton, the Newnes people, the UK people, the EDN people, and the people of Butterworth–Heinemann for having so much confidence in me. I recognize that all books are a team effort and am thankful that I now work with the New First Team on this series.

And to Irene, my wife and super agent, I extend my thanks. Without her help, this book could not have been written.

CHAPTER 1

Voltage-Frequency Converter Basics

This chapter is devoted to basic voltage-frequency converters, both voltage-to-frequency and frequency-to-voltage. It is primarily for readers who are totally unfamiliar with voltage-frequency conversion. It is possible to design voltage-frequency circuits "from scratch." However, voltage-frequency converters are available in integrated circuit (IC) form, and it is generally simpler to use such ICs.

The data sheets for IC converters often show the connections and provide all necessary design parameters to produce complete converter circuits by adding external components. This chapter describes the functions and operation of IC converters to help you understand the data-sheet information.

Before we get started, let us resolve certain differences in terms. Some manufacturers refer to voltage-to-frequency converters as VFCs. Other manufacturers use the term V/F converter. The same is true for frequency-to-voltage converters, which are referred to as FVCs by some, and as F/V converters by others. We prefer the terms VFC and FVC, but do not be surprised to find both terms in this book.

1.1 Basic Voltage-Frequency Conversion Techniques

This section describes the various VFC and FVC techniques in common use. Here, we concentrate on explanations of the basic principles used in voltage-frequency conversion. By studying this information, you should be able to understand operation of the converter ICs that are described throughout the book. It is assumed that you are familiar with the principles of radio-frequency (RF) transmission and with basic digital electronics. If not, read the author's *Lenk's RF Handbook* (McGraw-Hill, 1992) and *Lenk's Digital Handbook* (McGraw-Hill, 1993).

1.1.1 Basic VFC Operation

Most present-day VFC and FVC circuits use a VFC IC as the basic converter element. Figure 1–1 shows such an IC connected as a VFC. The circuit is essentially a relaxation oscillator with an output frequency proportional to input voltage. The voltage to be converted is applied to a comparator within the IC at pin 7. The comparator output is applied to a one-shot. Output pulses from the one-shot are applied to pin 3 (the circuit output) through Q1 and to a current switch.

The current switch interrupts currents to R_L and it thus provides current pulses at pins 1 and 6. Except at zero, the current pulses keep the average voltage across C_L (at pin 6 of the comparator) slightly greater than the input voltage, so the one-shot continues to produce pulses. Resistor R_S is made adjustable so that the current source can be set to a given scale factor (frequency out for a given voltage in), with R_T and C_T selected for some given frequency range.

1.1.2 Basic FVC Operation

The same VFC shown in Fig. 1–1 can also be used for frequency-to-voltage conversion. Figure 1–2 shows typical connections. Note that for FVC, the frequency input (pulse or square wave) is applied to the comparator at pin 6. This connection produces pulses at the comparator output, which correspond in frequency to the input.

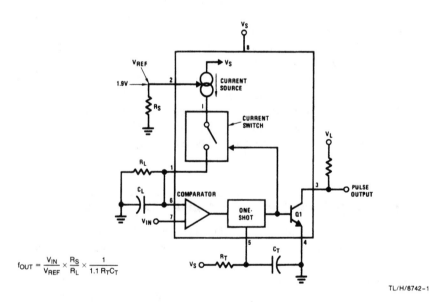

$$f_{OUT} = \frac{V_{IN}}{V_{REF}} \times \frac{R_S}{R_L} \times \frac{1}{1.1\,R_T C_T}$$

TL/H/8742–1

Figure 1–1. VFC IC connected as a VFC (National Semiconductor, *Linear Applications Handbook*, 1994, p. 1247)

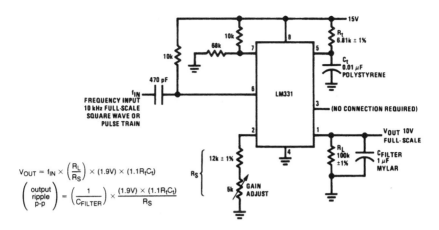

Figure 1–2. VFC IC connected as an FVC (National Semiconductor, *Linear Applications Handbook,* 1994, p. 1204)

The comparator pulses trigger the one-shot. In turn, the one-shot controls the amount of current at pin 1 (and thus the circuit output voltage) through the current switch. Resistor R_S is made adjustable so that the current source can be set to produce a given scale factor (voltage out for a given frequency in), with R_T and C_T selected for some given frequency range. In the circuit of Fig. 1–2, R_S is adjusted so that the voltage output (pin 1) is 10 V when the input frequency is 10 kHz.

1.1.3 Discrete-Component VFC Operation

VFC ICs cannot be used for all voltage-to-frequency applications, so a number of discrete-component circuits have been developed to meet special needs. The remainder of this section is a summary of discrete-component VFC circuit techniques. Chapter 6 describes a number of discrete-component VFC and FVC circuit approaches.

1.1.4 Ramp-Comparator VFC

Figure 1–3 shows the basic ramp-comparator VFC concept. The input drives an integrator, and the slope of the integrator ramp varies with the input-derived current. When the ramp crosses V_{REF}, the comparator turns on the switch, discharging the capacitor. This restarts the cycle. The frequency of this action directly relates to input voltage. In some designs, one op amp serves as both integrator and comparator.

1.1.5 Charge-Pump VFC

In the circuit of Fig. 1–4, the integrator is enclosed in a charge-dispensing loop. Capacitor C1 charges to V_{REF} during the integrator-ramp time. When the comparator

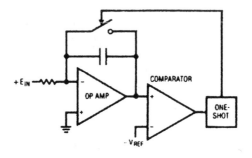

Figure 1–3.
Basic ramp-comparator
VFC concept (Linear
Technology, *Linear Appli-
cations Handbook,*1993,
p. AN3-7)

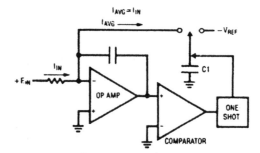

Figure 1–4.
Basic charge-pump VFC
(Linear Technology, *Lin-
ear Applications Hand-
book,* 1993, p. AN3-7)

trips, C1 is discharged into the op-amp summing point, forcing the op amp high. After C1 discharges, the op amp begins to ramp and the cycle repeats (frequency is related to input voltage).

1.2 Voltage-Frequency Converter Terms and Data Sheets

Much of the basic design information for a particular IC voltage-frequency
converter can be obtained from the data sheet. Likewise, a typical data sheet describes a few specific applications for the converter. However, converter data sheets
often have two weak points. First, they assume that everyone understands all of the
terms used. Of more importance, the data sheets do not show how the listed parameters relate to design problems. To further complicate the situation, each manufacturer
has a separate system of data sheets. It is impractical to discuss all data sheets here.
Instead, we discuss typical information found on the converter data sheets, and see
how this information affects simplified design.

1.2.1 Typical Data-Sheet Information

Figure 1–5 shows typical IC voltage-frequency converters (the Raytheon
4151/4152, which are discussed in greater detail throughout Chapter 2). Figures 1–6

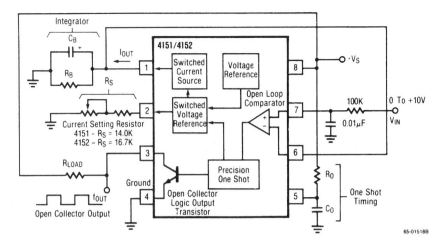

Figure 1-5. Typical IC voltage-frequency converter (*Raytheon Semiconductor Data Book,* 1994, p. 7-6)

through 1–8 show typical data-sheet information for the IC, including pinouts, maximum ratings, thermal characteristics, electrical characteristics, and typical performance characteristics.

1.2.2 IC Voltage-Frequency Converter Operation

Before we cover the definitions of terms, let us review how the basic converter IC operates. As shown in Fig. 1–5, the IC contains an open-loop comparator, a precision one-shot timer, a switched voltage reference, a switched current source, and an open-collector logic-output transistor. The basic IC is converted to a single-supply VFC by adding a few external resistors and capacitors.

The comparator output controls the one-shot (which is actually a monostable timer). In turn, the one-shot controls the switched current source, the switched reference, and the open-collector transistor. If the voltage at pin 7 is greater than the voltage at pin 6, the comparator switches and triggers the one-shot. When the one-shot is triggered, the timing period begins, and the switched current source, switched voltage reference, and transistor are turned on.

The one-shot creates its timing period much like the classic 555 timer, by charging a capacitor from a resistor tied to $+V_S$. The one-shot senses the voltage on the capacitor (pin 5) and ends the timing period when the voltage reaches two-thirds of the supply voltage. The capacitor is discharged by the transistor at the end of the timing period.

During the timing period, the current source, the switched reference, and the output transistor are switched on. The switched current source (pin 1) delivers a current proportional to both the reference voltage and the external resistor R_S. The

Connection Information

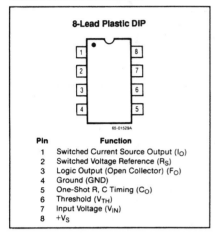

8-Lead Plastic DIP

65-01529A

Pin	Function
1	Switched Current Source Output (I_O)
2	Switched Voltage Reference (R_S)
3	Logic Output (Open Collector) (F_O)
4	Ground (GND)
5	One-Shot R, C Timing (C_O)
6	Threshold (V_{TH})
7	Input Voltage (V_{IN})
8	$+V_S$

Functional Block Diagram

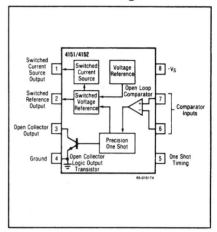

65-01517A

Absolute Maximum Ratings

Supply Voltage .... +22V
Internal Power Dissipation .... 500 mW
Input Voltage .... -0.2V to $+V_s$
Output Sink Current
    (Frequency Output) .... 20 mA
Output Short Circuit to Ground .... Continuous
Storage Temperature
    Range .... -65°C to +150°C
Operating Temperature
    Range .... 0°C to +70°C

Ordering Information

Part Number	Package	Operating Temperature Range
RC4151N	N	0°C to +70°C
RC4152N	N	0°C to +70°C

Notes:
N = 8- lead plastic DIP
Contact a Raytheon sales office or representative for ordering information on special package/temperature range combinations.

Thermal Characteristics

	8-Lead Plastic DIP
Max. Junction Temp.	125°C
Max. P_D T_A <50°C	468 mW
Therm. Res θ_{JC}	—
Therm. Res. θ_{JA}	160°C/W
For T_A >50°C Derate at	6.25 mW/°C

Figure 1–6. Typical VFC data sheet information (*Raytheon Semiconductor Data Book,* 1994, p. 7-3)

switched reference (pin 2) supplies an output voltage equal to the internal reference voltage (4151 = 1.9 V, and 4152 = 2.25 V, as shown in Fig. 1–7). The output transistor is turned on, forcing the logic output (pin 3) to a low state.

At the end of the timing period, all of these outputs are turned off. As a result, the switched voltage reference has produced an off-on-off voltage pulse, the switched

Electrical Characteristics (V_S = +15V and T_A = +25°C unless otherwise noted)

Parameters	Test Conditions	4151			4152			Units
		Min	Typ	Max	Min	Typ	Max	
Power Supply Requirements (Pin 8) Supply Current	V_S = +15V		4.5	7.5		2.5	6.0	mA
Supply Voltage		+8.0	+15	+22	+7.0	+15	+18	V
Input Comparator (Pins 6 and 7) V_{OS}			±2.0	±10		±2.0	±10	mV
Input Bias Current			-100	-300		-50	-300	nA
Input Offset Current			±50	±100		±30	±100	nA
Input Voltage Range		0	V_S-2	V_S-3	0	V_S-2	V_S-3	V
One Shot (Pin 5) Threshold Voltage		0.63	0.67	0.70	0.65	0.67	0.69	XV_S
Input Bias Current			-100	-500		-50	-500	nA
Saturation Voltage	I = 2.2mA		0.15	0.5		0.1	0.5	V
Drift of Timing vs. Temperature[2]	T = 75µS 0°C to +70°C		±35			±30	±50	ppm/°C
Drift of Timing vs. Supply			±150			±100		ppm/V
Switched Current Source[1] (Pin 1) Output Current	4151-R_S = 14.0K/ 4152-R_S = 16.7K		+138			+138		µA
Drift vs. Temperature[2]	0°C to +70°C		±75			±50	±100	ppm/°C
Drift vs. Supply Voltage			0.15			0.10		%/V
Leakage Current	Off State		1.0	50		1.0	50	nA
Compliance	Pin 1 = 0V to +10V	1.0	2.5		1.0	2.5		µA
Reference Voltage (Pin 2) V_{REF}		1.7	1.9	2.08	2.0	2.25	2.5	V
Drift vs. Temperature[2]	0°C to +70°C		±50			±50	±100	ppm/°C
Logic Output (Pin 3) Saturation Voltage	I_{SINK} = 3.0mA		0.1	0.5		0.1	0.5	V
Saturation Voltage	I_{SINK} = 10mA		0.8			0.8		V
Leakage Current	Off State		0.2	1.0		0.1	1.0	µA
Nonlinearity % Error Voltage Sourced Circuit of Figure 3	1.0Hz to 10kHz		0.013			0.007	0.05	%
Temperature Drift Voltage[2] Sourced Circuit of Figure 3	0°C to +70°C F_0 = 10kHz		±100			±75	±150	ppm/°C

Notes:
1. Temperature coefficient of output current source (pin 1 output) exclusive of reference voltage drift.
2. Guaranteed but not tested.

Figure 1-7. Typical VFC electrical characteristics (*Raytheon Semiconductor Data Book,* 1994, p. 7-4)

current source has emitted a given charge to the comparator and integrator, and the transistor has produced an output logic pulse.

As a point of reference, the values of C_O and C_B should be 0.1 µF and 10 µF for an operating frequency of DC to 1 kHz, 0.01 µF and 1 µF for DC to 10 kHz, and 0.001 µF and 0.1 µF for DC to 100 kHz. The values of R_O and R_B should be 6.8 k and

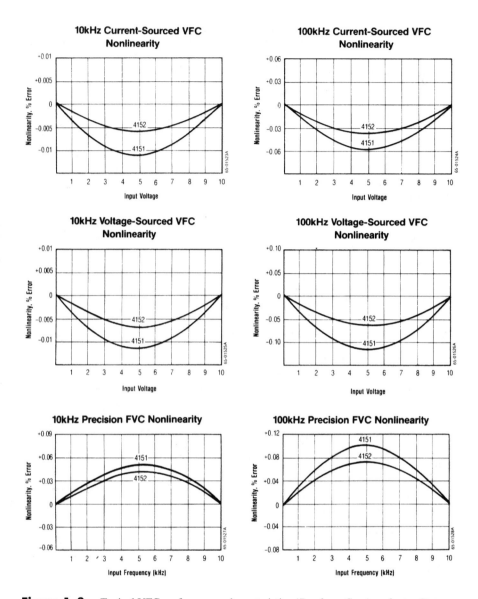

Figure 1–8. Typical VFC performance characteristics (*Raytheon Semiconductor Data Book,* 1994, p. 7-5)

100 k for all operating frequencies, between DC and 100 kHz. Chapter 2 describes the simplified design procedures for external component selection in much greater detail.

1.2.3 Voltage-Frequency Converter Definitions

The remaining paragraphs in this section describe characteristics that are generally the most important in design.

1.2.4 Compliance

Compliance is the measure of output impedance of a switched-current source, given as a maximum current for a specified voltage change. As shown in Fig. 1–7, compliance for the 4151/4152 goes from a minimum of 1 μA to a typical 2.5 μA, with a pin-1 voltage from 0 V to +10 V.

1.2.5 Full-Scale Frequency

A VFC can operate up to the guaranteed full-scale frequency without violating any of the performance specifications for this frequency range. Although no frequency limit is given for the 4151/4152, the performance characteristics of Fig. 1–8 show operation at a frequency of 100 kHz.

1.2.6 Nonlinearity Error

On a plot of input voltage versus output frequency (Fig. 1–8), a straight line is drawn from the origin to the full-scale point, which is defined by the intersection of the maximum input voltage and maximum output frequency. The actual plot of output frequency F_O versus input voltage should not deviate from this straight line by more than the increment ΔF_O (max).

Nonlinearity is defined here as $(\Delta F_O/\Delta F_S) \times 100\%$ where F_S is the maximum frequency for the range in question. For example, when specifying nonlinearity error for the 0.1-Hz to 10-kHz range, then $F_S = 10$ kHz. When specifying nonlinearity error for a FVC, nonlinearity error is defined as $(\Delta V/V_{FS}) \times 100\%$.

1.2.7 Leakage Current

Leakage current is the current that flows into the open-collector output transistor when the transistor is in the "off" state, as a result of maximum supply voltage being applied to the output. As shown in Fig. 1–7, leakage current for the 4151/4152 goes from a typical 1 nA to a maximum of 50 nA, in the "off" state.

1.2.8 Reference Voltage

Reference voltage or V_{REF} is the output of the internal voltage reference. This cannot be measured directly in the 4151/4152, but can be measured as the switched voltage at pin 2 (Fig. 1–5). In some VFCs, the internal reference voltage is accessible at one of the IC pins (such as at pin 5 of the 4153 discussed in Chapter 2). When V_{REF} can be measured externally, another characteristic (reference current) is sometimes given on the data sheet. Reference current is the current produced (and usually returned back to circuits in the IC) when the internal voltage reference is at some exact value (such as 7.3 V for the 4153). As shown in Fig. 1–7, V_{REF} for the 4151 goes from a minimum of 1.7 V to a maximum of 2.08 V. V_{REF} for the 4152 goes from 2.0 V to 2.5 V.

1.2.9 Scale Factor

Scale factor K is the ratio of output frequency divided by voltage in, or F_O/V_{IN}. No scale factor is given directly for the 4151/4152. However, a scale-factor tolerance is given for the 4153, and involves V_{REF}, resistance in (R_{IN}), and capacitance out (C_O), where $K = 1/2V_{REF} R_{IN} C_O$. The scale-factor tolerance for the 4153 is a typical ±0.5% at a frequency of 10 kHz. From a simplified-design standpoint, the change in scale factor with changes in supply voltage and/or temperature is of most importance.

1.2.10 Absolute Maximum Ratings and Thermal Characteristics

The maximum ratings and thermal characteristics shown in Fig. 1–6 deal primarily with the IC rather than the internal circuits. These characteristics are important to design when selecting heat sinks (if any) and when choosing ICs for a particular operating environment. If you are not familiar with such IC characteristics, read the author's *Simplified Design of Linear Power Supplies* (1994) and *Simplifed Design of Switching Power Supplies* (1995), both published by Butterworth–Heinemann.

1.3 Basic VFC Tests

In simplified design, the obvious test for any VFC is to vary the input voltage over the range and check that the output frequency varies accordingly. Use a digital meter at the input and a frequency counter at the output. Using the discrete-component circuit of Fig. 1–9 as an example, the output frequency should vary between 0 and 30 kHz when the input voltage is varied between 0 and 3 V (1-V input produces 10-kHz output, 2-V input produces 20-kHz output, and so on). If practical, the circuit can be subjected to temperature changes and the output frequency monitored for drift. With the circuit of Fig. 1–9, the drift is supposed to be about 20 ppm/°C.

Using the circuit of Fig. 1–10 as an example, two inputs are required for testing, and the output should be the ratio of the two inputs. That is, if V1 is –10 V and 2 is –8 V, the ratio is 10/8 = 1.25, and the output should be 12.5 kHz. Note that full-scale for the circuit of Fig. 1–10 is 15 kHz, so ratios beyond 1.5 cannot be measured.

1.4 Basic FVC Tests

The test for an FVC is the reverse of that for a VFC. That is, you vary the input frequency over the range and check that output voltages vary accordingly. Generally, a pulse generator or possibly a square-wave generator is recommended for the input.

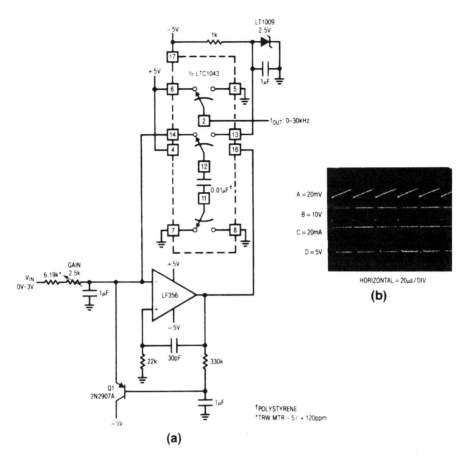

Figure 1-9. Charge-pump VFC (Linear Technology, *Linear Applications Handbook,* 1993, p. AN3-11)

In some cases, you can use a simple RC differentiator to convert from sine waves to pulses at the input, but this can disturb operation of some circuits. As a general guideline, use the sine-wave and RC differentiator combination for the input only when a pulse or square-wave generator is not available, or if a sine-wave is specifically recommended in the data sheet.

1.5 Basic VFC Troubleshooting

The first step in troubleshooting VFC circuits involves checking that the desired output frequency is produced by a given input voltage. If the output is not correct, try correcting the problem with adjustment. If the problem cannot be corrected

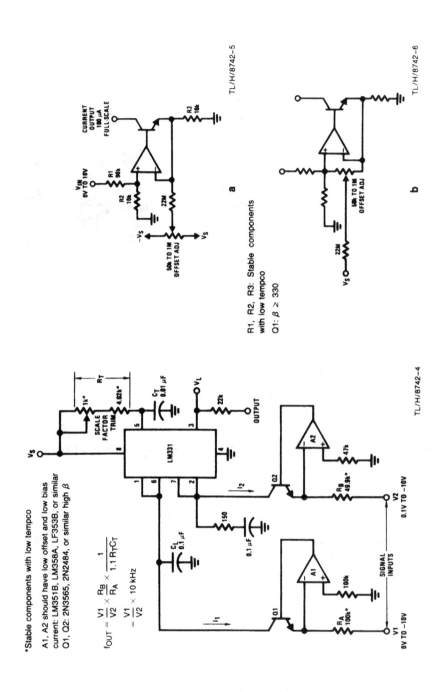

Figure 1-10.
Ratio VFC (National Semiconductor, *Linear Applications Handbook*, 1994, p. 1249)

by adjustment, trace signals using a meter or scope from input (typically a DC voltage) to output (typically pulses). From that point on, it is a matter of voltage measurements and/or point-to-point resistance measurements. The following are typical examples.

In the circuit of Fig. 1–9, begin by checking for pulses at pin 2 of the LTC1043. Pin 2 switches between pins 5 (ground) and 6 (+5 V) at a frequency determined by the signal at clock pin 16. If there is no signal at pin 16, the LTC switches at a frequency rate near 200 kHz. In this circuit, the LTC1043 clock is synchronized with the signal at pin 16.

If pin 2 is not switching at any frequency, suspect the LTC1043. If pin 2 is switching at a fixed frequency, with a variable V_{IN}, suspect the LF356 and associated parts. The output of the LF356, trace B in Fig. 1–9b, should be a series of negative pulses. The inverting input (summing point) is a series of positive ramps (trace A). Current flowing from the LF356 summing point into the 0.01-μF capacitor at the end of the ramp should produce a series of negative spikes (trace C). Simultaneously, there should be a series of pulses (trace D) at the noninverting input of the LF356.

Note that Q1 prevents the LF356 from going to the negative rail (and staying there) by pulling the summing point negative if the output stays low long enough to charge the 1-μF/330-k RC during startup. Also note that if the circuit shows excessive drift, or nonlinearity in output frequency (for a given input voltage), suspect the 0.01-μF capacitor.

To trim the circuit of Fig. 1–9, apply 3.0 V, and adjust gain trim for a 30-kHz output. The output frequency is directly related to input voltage, with a supposed transfer linearity of 0.005%.

To troubleshoot the circuit of Fig. 1–10, start by checking for pulses at the 22-k resistor (pin 3 of LM331). There should be pulses at this output regardless of what voltages are applied at the input (including zero input). With both V1 and V2 at zero, the scale-factor trim is adjusted so that the output is 10 kHz.

The circuit of Fig. 1–10 converts the ratio of two voltages to an equivalent frequency. The two op amps convert the inputs to proportional currents. The 1-k scale-factor trim is adjusted so that the frequency output equals the ratio of V1/V2 × 10 kHz. Full-scale output is 15 kHz.

The circuit can accept positive inputs when the op amps are rearranged, as shown in Figs. 1–10a or 1–10b. Trimming out the offset in the op amp gives the ratio converter better linearity and accuracy. The trim circuit in Fig. 1–10a needs stable positive and negative supplies for the offset trimmer, but the trim in Fig. 1–10b needs only an stable positive supply. Unmarked components in Fig. 1–10b are the same as those in Fig. 1–10a.

If there are no pulses at pin 3 of the LM331, suspect the LM331, or possibly the timing capacitor C_T. The same is true if there are pulses, but the frequency cannot be brought within the desired range, or the frequency does not change with changing voltage at the input (pins 1/6 and/or 2/7).

If there are pulses, but the pulse frequency is not controlled by voltages V1 and V2, suspect Q1, Q2, A1, A2, and the associated parts. Make certain that the voltages at pins 1/6 and 1/7 vary when V1 and V2 are varied.

1.6 Basic FVC Troubleshooting

The first step in troubleshooting FVC circuits involves checking that the desired output voltage is produced by a given input frequency. If the output is not correct, try correcting the problem with adjustment. If the problem cannot be corrected by adjustment, trace signals using a meter or scope from input (typically pulse or square-wave signals) to the output (typically a DC voltage). After tracing, make voltage measurements and/or a point-to-point resistance measurement. The following are typical examples.

In the circuit of Fig. 1–11, the frequency input is applied to the clock input of an LTC1043 switched-capacitor IC. The 1000-pF capacitor is switched between a fixed voltage and the inverting input of the LF356. The 1-µF feedback capacitor averages this action over several cycles, and the circuit output is a DC level that is linearly related to frequency. The feedback resistors set the LF356 gain. Note that the input pulse must be low for at least 100 ns to allow complete discharge of the 1000-

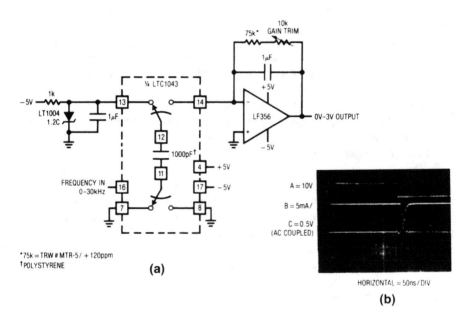

Figure 1–11. Charge-pump FVC (Linear Technology, *Linear Applications Handbook,* 1993, p. AN3-12)

µF capacitor. To trim the circuit, apply 30-kHz to pin 16 of the LTC1043 and set the gain trim for exactly 3.0 V.

To troubleshoot the Fig. 1–11 circuit, start by checking for pulses at pin 14 of the LTC1043. Pin 12 switches between pins 13 and 14 at a frequency that is determined by the signal at clock pin 16 (trace A of Fig. 1–11b), which is the FVC-circuit input in this case.

If pin 14 is not switching at any frequency, suspect the LTC1043 (of course, check for proper voltages at pins 4, 13, and 17, as well as ground at pins 7 and 8). If the LTC1043 and voltages are good, but there are no pulses at pin 14, the 1000-pF capacitor at pins 11 and 12 is the prime suspect (trace B shows the capacitor signal). The capacitor might be shorted or badly leaking.

If there are pulses at the inverting input of the LF356, but there is no output voltage, or if the output voltage does not vary with changes in pulse frequency at the input, suspect the LF356 or associated parts. The feedback resistors determine the LF356 gain, and the 1-µF feedback capacitor averages the pulse input to a DC output (trace C shows the negative and positive swing of the LF356 output).

Note that if the circuit shows excessive drift, or nonlinearity in output voltage (for a given input frequency), suspect the 1000-pF capacitor at pins 11 and 12.

To troubleshoot the circuit of Fig. 1–2 (Section 1.1.2), start by checking for a DC voltage at pin 1 of the LM331, with a signal at the input. (If the input to the LM331 is zero, the voltage at pin 1 is zero.) If the voltage does not vary at pin 1, when the frequency of the signal at pin 6 is varied (between 0 and 10 kHz), suspect the LM331. It is also possible that timing capacitor C_T is shorted or badly leaking.

Try monitoring the output voltage (with the input frequency steady) while varying the gain-adjust control. If there is no change in output voltage, check voltage and resistance from pin 2 of the LM331 to ground. If the connection from pin 2 is good, suspect the LM331.

CHAPTER 2

Simplified Design with Raytheon VFCs

This chapter is devoted to simplified-design approaches for Raytheon VFCs (the 4151, 4152, and 4153 discussed in Chapter 1). All of the general design information in Chapter 1 applies to the examples in this chapter. However, each voltage-frequency IC has special design requirements, all of which are discussed in detail. The circuits in this chapter can be used immediately the way they are, or by altering component values, as a basis for simplified design of similar voltage-frequency conversion applications. (Note that Raytheon converter ICs carry the designations RC4151, RC4152, and RC4153.) The first sections of the chapter describe operation and characteristics of the ICs. The remaining sections describe using the ICs in specific applications.

2.1 Description of 4151/4152

The characteristics and basic operation of the 4151/4152 are given in Chapter 1. Figure 1–5 shows the IC connected for basic single-supply operation. Section 1.2.2 describes how the IC works. This information will not be repeated here. Instead, we will go directly into how the IC can be connected to provide various VFC and FVC functions.

2.1.1 Precision Current-Sourced VFC (RC4152)

Figure 2–1 shows the IC connected as a precision current-sourced VFC. This is similar to the single-supply VFC (Fig. 1–5) except that the passive RC integrator is replaced by an active op-amp integrator. The op amp increases the dynamic range down to 0 V, improves the response time, and eliminates the nonlinearity error introduced by the limited compliance of the switched current-source output. The integrator algebraically sums the positive current pulses from the switched current source

17

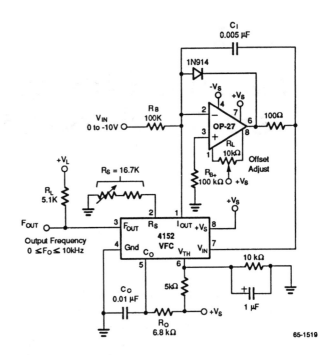

Figure 2-1.
Precision current-sourced
VFC (*Raytheon Semicon-
ductor Data Book,* 1994,
p. 3-805)

with the current produced by V_{IN}/RB. To operate correctly, the input voltage must be negative, so that when the circuit is balanced, the two currents cancel.

Figure 2–2 shows recommended component values for different operating frequencies. Figure 2–3 shows nonlinearity versus input voltage for the precision current-sourced VFC. Compare this to the performance characteristics shown in Fig. 1–8. For best results, use an op amp having a slew rate of 1 V/μs or greater.

2.1.2 Precision Voltage-Sourced VFC (RC4152)

Figure 2–4 shows the IC connected as a precision voltage-sourced VFC. The circuit is identical to the current-sourced VFC, except that the current pulses applied to the integrator are taken directly from the switched voltage reference. This im-

Range		Scale				
Input V_{IN}	Output F_O	Factor	R_O	C_O	C_I	R_B
0 to -10V	0 to 1.0 kHz	0.1 KHz/V	6.8 kΩ	0.1 μF	0.05 μF	100 kΩ
0 to -10V	0 to 10 kHz	1.0 kHz	6.8 kΩ	0.01 μF	0.005 μF	100 kΩ
0 to -10V	0 to 100 kHz	10 kHz/V	6.8 kΩ	0.001 μF	500 pF	100 kΩ

Figure 2–2. Recommended component values for different operating frequencies
(*Raytheon Semiconductor Data Book,* 1994, p. 3-804)

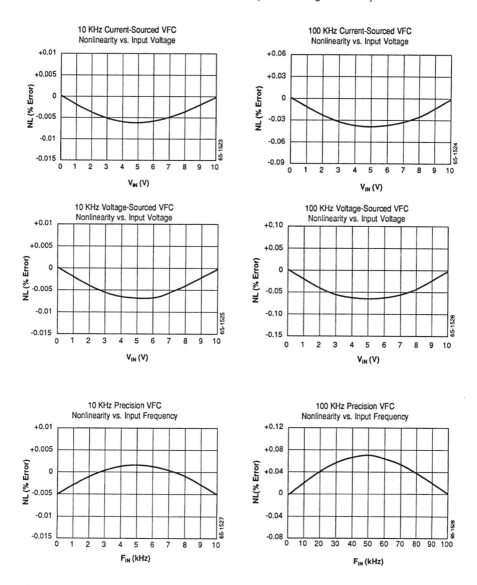

Figure 2–3. Typical performance characteristics for 4152 (*Raytheon Semiconductor Data Book,* 1994, p. 3-802)

proves temperature drift (at the expense of high-frequency linearity), as shown in Fig. 2–3.

2.1.3 Single-Supply FVC (RC4152)

Figure 2–5 shows the IC connected as a single-supply FVC. The circuit converts an input pulse train into an average output voltage. Incoming pulses trigger the

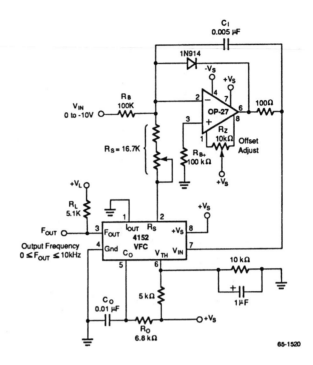

Figure 2-4.
Precision voltage-sourced
VFC (*Raytheon Semicon-*
ductor Data Book, 1994,
p. 3-805)

$$V_{OUT} = \left[\frac{1.1\ R_oC_oR_B\ V_{REF}}{R_s} \right] F_{IN}\ (Hz)$$

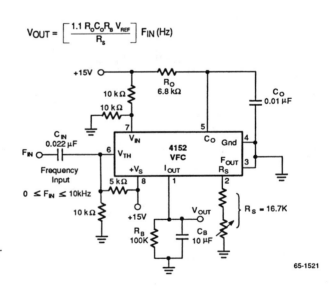

Figure 2-5.
Single-supply FVC
(*Raytheon Semiconductor*
Data Book, 1994,
p. 3-806)

Input Operating Range	C_{IN}	R_O	C_O	R_B	C_B	Ripple
0 to 1.0 kHz	0.02 µF	6.8 kΩ	0.1 µF	100 kΩ	100 µF	1.0 mV
0 to 10 kHz	0.002 µF	6.8 kΩ	0.01 µF	100 kΩ	10 µF	1.0 mV
0 to 100 kHz	200 pF	6.8 kΩ	0.001 µF	100 kΩ	1.0 µF	1.0 mV

Figure 2–6. Recommended values for various operating ranges (*Raytheon Semiconductor Data Book,* 1994, p. 3-806)

input comparator and fire the one-shot, which dumps a charge into the output integrator. The voltage on the integrator becomes a varying DC voltage that is proportional to the input signal frequency.

The input waveform must have fast-slewing edges, and the differentiated input signal must be less than the timing period of the one-shot (1.1 $R_O C_O$). Figure 2–6 shows recommended values for various operating ranges.

The differentiator and divider are used to shape and bias the trigger input. A negative-going pulse at pin 6 causes the comparator within the IC to fire the one-shot. The input pulse amplitude must be large enough to trip the comparator, but not so large as to exceed the IC's input-voltage ratings.

Output voltage is directly proportional to input frequency (0 to 10 kHz, 0 to 10 V). Minor adjustments to linearity can be made by varying the value of R_S.

2.1.4 Precision FVC (RC4152)

Figure 2–7 shows the IC connected as a precision FVC, where an op amp has been added to improve linearity, offset, and response time. The op-amp circuit acts as an active lowpass single-pole filter. Response time can be further improved by adding a double-pole filter in place of the single-pole circuit shown. The bottom two graphs in Fig. 2–3 show nonlinearity error versus input frequency (even though they are labeled as a VFC!). Again, minor adjustments to linearity can be made by varying the value of R_S. Adjust R_Z to provide the desired offset. Use 0.005 µF as a first trial value for C1.

2.2 Description of 4153

Figure 2–8 shows the connection information, maximum ratings, and thermal characteristics for the Raytheon RC4153 converter. Figures 2–9 and 2–10 show the electrical characteristics and typical performance characteristics, respectively. Figures 2–11, 2–12, and 2–13 show the IC connected as a minimum-circuit VFC, FVC, and VFC with offset and gain adjustments, respectively. The circuits of Figs. 2–11

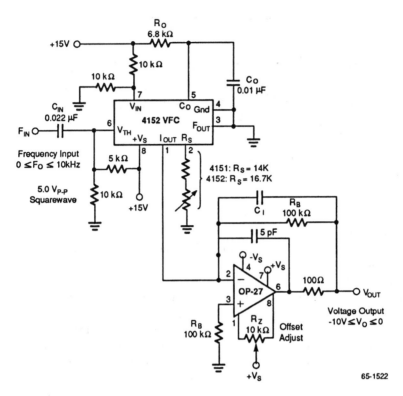

Figure 2–7. Precision FVC (*Raytheon Semiconductor Data Book,* 1994, p. 3-807)

through 2–13 can be used directly, as is, for simplified design. However, the following paragraphs describe both operating principles and detailed circuit operation. A study of these paragraphs may help the reader to understand the applications given in the remainder of this chapter.

2.2.1 Operating Principles (VFC)

Figure 2–14 shows a block diagram of the IC connected for VFC operation. Figure 2–15 shows the corresponding waveforms and timing.

Both capacitors shown in Fig. 2–14 are discharged when power is first applied. The input current, produced by V_{IN}/R_{IN}, causes C1 to charge, and a ramp to be developed at point C. The trigger threshold of the one-shot is about +1.3 V and, if the integrator output (point C) is less than +1.3 V, the one-shot will fire and pulse the open-collector output (point E) and the switched current-source output (point A).

Because point C is less than +1.3 V, the one-shot fires, and the switched current source delivers a negative current pulse to the integrator. This causes C1 to to charge in the opposite direction, and point C ramps up until the end of the one-shot pulse. At

Absolute Maximum Ratings(1)

Supply Voltage ....±18V
Internal Power Dissipation ....500 mW
Input Voltage ....-V$_S$ to +V$_S$
Output Sink Current
(Frequency Output) ....20 mA
Storage Temperature Range ....-65°C to +150°C
Operating Temperature Range
RM4153 ....-55°C to +125°C
RC4153...0°C to +70°C

Note:
1. "Absolute maximum ratings" are those beyond which the
    safety of the device cannot be guaranteed. They are not meant
    to imply that the device should be operated at these limits. If
    the device is subjected to the limits in the absolute maximum
    ratings for extended periods, its reliability may be impaired.
    The tables of Electrical Characteristics provide conditions for
    actual device operation.

Thermal Characteristics

	14-Lead Ceramic DIP
Max. Junction Temp.	+175°C
Max. P$_D$ T$_A$ <50°C	1042 mW
Therm. Res θ$_{JC}$	60° C/W
Therm. Res. θ$_{JA}$	120°C/W
For T$_A$ >50°C Derate at	8.33 mW/°C

Ordering Information

Part Number	Package	Operating Temperature Range
RC4153D	D	0°C to +70°C
RM4153D	D	-55°C to +125°C

Notes:
D = 14-lead ceramic DIP

Connection Information

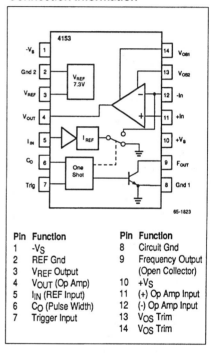

Pin	Function	Pin	Function
1	-V$_S$	8	Circuit Gnd
2	REF Gnd	9	Frequency Output
3	V$_{REF}$ Output		(Open Collector)
4	V$_{OUT}$ (Op Amp)	10	+V$_S$
5	I$_{IN}$ (REF Input)	11	(+) Op Amp Input
6	C$_O$ (Pulse Width)	12	(-) Op Amp Input
7	Trigger Input	13	V$_{OS}$ Trim
		14	V$_{OS}$ Trim

Figure 2–8. Connection information, maximum ratings, and thermal characteristics for RC4153 (*Raytheon Semiconductor Data Book,* 1994, p. 3-810)

that time, the positive current V$_{IN}$/R$_{IN}$ again makes point C ramp down until the trigger threshold is reached.

The one-shot continuously fires until the integrator output exceeds the trigger threshold. When this point is reached, the one-shot fires as needed to keep the integrator output above the trigger threshold. If V$_{IN}$ is increased, the slope of the down ramp increases, and the one-shot fires more often to keep the integrator output high.

Because the one-shot firing frequency is the same as the open-collector output frequency, any increase in V_{IN} causes an increase in F_{OUT} (shown in Figs. 2–11 through 2–13). This relationship is almost linear because the amount of charge in each output current pulse is carefully defined, both in magnitude and duration. The pulse duration is set by the timing capacitor C_O (point D). This feedback system is called a charge-balanced loop.

Electrical Characteristics

($V_S = \pm15V$ and $T_A = +25°C$ unless otherwise noted)

Parameters	Min	Typ	Max	Units
Power Supply Requirements				
Supply Voltage	±12	±15	±18	V
Supply Current $(+V_S, I_{OUT} = 0)$		+4.2	+7.5	mA
$(-V_S, I_{OUT} = 0)$		-7	-10	
Full Scale Frequency	250	500		kHz
Transfer Characteristics				
Nonlinearity Error Voltage-to-Frequency[1]				
$0.1\,Hz \le F_{OUT} \le 10\,kHz$		0.002	0.01	%FS
$1.0\,Hz \le F_{OUT} \le 100\,kHz$		0.025	0.05	%FS
$5.0\,Hz \le F_{OUT} \le 250\,kHz$		0.06	0.1	%FS
Nonlinearity Error Frequency-to-Voltage[1]				
$0.1\,Hz \le F_{IN} \le 10\,kHz$		0.002	0.01	%FS
$1.0\,Hz \le F_{IN} \le 100\,kHz$		0.05	0.1	%FS
$5.0\,Hz \le F_{IN} \le 250\,kHz$		0.07	0.12	%FS
Scale Factor Tolerance, F = 10 kHz $K = \dfrac{1}{2V_{REF}\,R_{IN}\,C_O}$		±0.5		%
Change of Scale Factor with Supply		0.008		%/V
Reference Voltage (V_{REF})		7.3		V
Temperature Stability (0°C to 70°C) [1, 2, 3]				
Scale Factor 10 KHz Nominal		±75	±150	ppm/°C
Reference Voltage		±50	±100	ppm/°C
Scale Factor (External Ref) 10 KHz FS		±25	±50	ppm/°C
Scale Factor (External Ref) 100 KHz FS		±50	±100	ppm/°C
Scale Factor (External Ref) 250 KHz FS		±100	±150	ppm/°C

Notes:
1. Guaranteed but not tested.
2. V_{REF} Range: $6.6V \le V_{REF} \le 8.0V$.
3. Over the specified operating temperature range.

Figure 2–9. Electrical characteristics for RC4153 (*Raytheon Semiconductor Data Book,* 1994, pp. 3-811, 3-812)

The scale factor K (Fig. 2–9), which is the number of pulses per second for a specified V_{IN}, is adjusted by changing either R_{IN} (and thus input current) or by changing the amount of charge in each I_{OUT} pulse. Because the magnitude of I_{OUT} is fixed at 1 mA, the amount of charge is adjusted by changing C_O to alter the one-shot duration. (I_{OUT} can be also be adjusted by changing V_{REF}, but this requires altering the fixed 7.3-V reference.) The accuracy of the relationship between V_{IN} and F_{OUT} is affected by three major sources of error: temperature drift, nonlinearity, and offset.

The greatest source of drift in a typical application is in the timing capacitor C_O. Low TC (temperature coefficient) capacitors, such as silver mica and polystyrene, should be measured for drift, using a capacitance meter. (The manufacturer recommends wiring parallel capacitors composed of 70% silver mica and 30% polystyrene.)

The built-in voltage reference (7.3 V) can also be replaced with a fixed external reference, if temperature/drift is critical. The manufacturer recommends an LM199 that has an output of 6.9 V, with less than 10 ppm/°C drift.

Parameters	Min	Typ	Max	Units
Op Amp				
Open Loop Output Resistance		230		Ω
Short Circuit Current		25		mA
Gain Bandwidth Product 1	2.5	3.0		MHz
Slew Rate	0.5	2.0		V/µS
Output Voltage Swing ($R_L \geq 2K$)	0 to +10	-0.5 to +14.3		V
Input Bias Current		70	400	nA
Input Offset Voltage (Adjustable to 0)		0.5	5.0	mV
Input Offset Current		30	60	nA
Input Resistance (Differential Mode)		1.0		MΩ
Common Mode Rejection Ratio	75	100		dB
Power Supply Rejection Ratio	70	106		dB
Large Signal Voltage Gain	25	350		V/mV
Switched Current Source				
Reference Current (External Reference)		1.0		mA
Digital Input (Frequency-to-Voltage, Pin 7)				
Logic "0"			0.5	V
Logic "1"	2.0			V
Trigger Current		-50		µA
Logic Output (Open Collector)				
Saturation Voltage (Pin 9)				
I_{SINK} = 4 mA		0.15	0.4	V
I_{SINK} = 10 mA		0.4	1.0	V
I_{LEAK} (Off State)		150		nA

Notes:
1. Guaranteed but not tested.

Figure 2–9. Continued.

The major cause of nonlinearity in this VFC (and most others) is that there is some change in the exact amount of charge in each I_{OUT} pulse. When the frequency increases, internal stray capacitances and switching problems change the width and

Typical Performance Characteristics

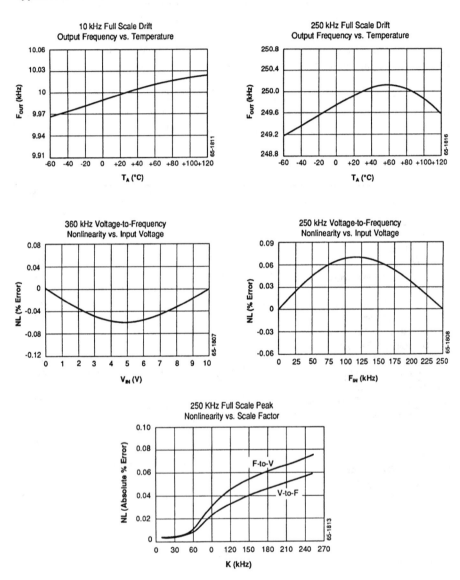

Figure 2-10. Typical performance characteristics for RC4153 (*Raytheon Semiconductor Data Book,* 1994, pp. 3-813, 3-814)

amplitude of the F_{OUT} pulses, causing a nonlinear relationship between V_{IN} and F_{OUT}. For this reason, the scale factor should be below 1kHz/V, or as low as the acquisition time of the system will permit.

Note that the circuits of Figs. 2–12 and 2–13 have an offset adjustment (between pins 13 and 14). This makes it possible to offset any drift in the internal op amp. As always, offset problems are most critical when the input voltages are low.

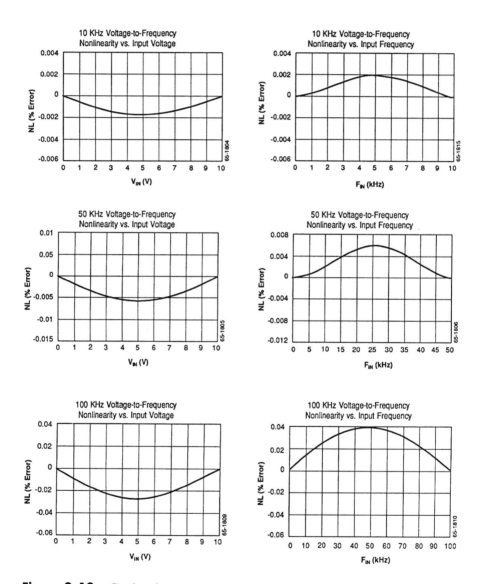

Figure 2–10. Continued.

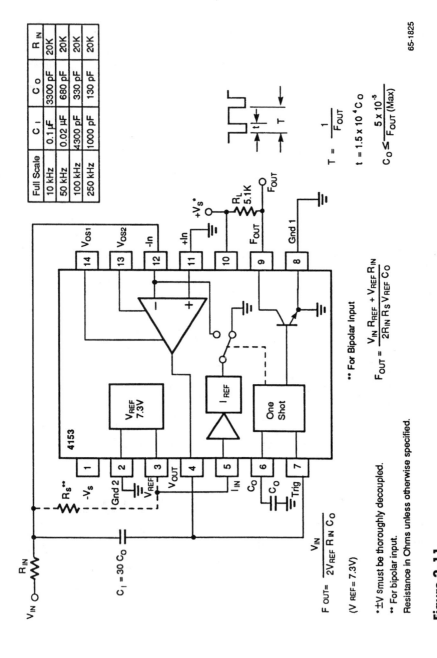

Full Scale	C_I	C_O	R_{IN}
10 kHz	0.1 µF	3300 pF	20K
50 kHz	0.02 µF	680 pF	20K
100 kHz	4300 pF	330 pF	20K
250 kHz	1000 pF	130 pF	20K

65-1825

$$T = \frac{1}{F_{OUT}}$$

$$t = 1.5 \times 10^{-4} C_O$$

$$C_O \leq \frac{5 \times 10^{-5}}{F_{OUT} \text{ (Max)}}$$

** For Bipolar Input

$$F_{OUT} = \frac{V_{IN} R_{REF} + V_{REF} R_{IN}}{2 R_{IN} R_S V_{REF} C_O}$$

$$F_{OUT} = \frac{V_{IN}}{2 V_{REF} R_{IN} C_O}$$

($V_{REF} = 7.3$V)

*±V s must be thoroughly decoupled.
** For bipolar input.
Resistance in Ohms unless otherwise specified.

Figure 2-11.
Minimum-circuit VFC (*Raytheon Semiconductor Data Book,* 1994, p. 3-815)

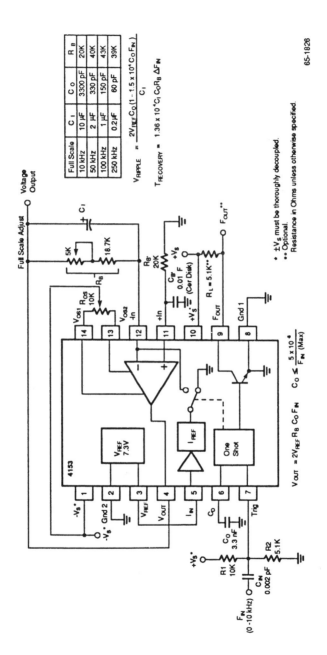

Figure 2-12.
Minimum-circuit FVC (*Raytheon Semiconductor Data Book*, 1994, p. 3-815)

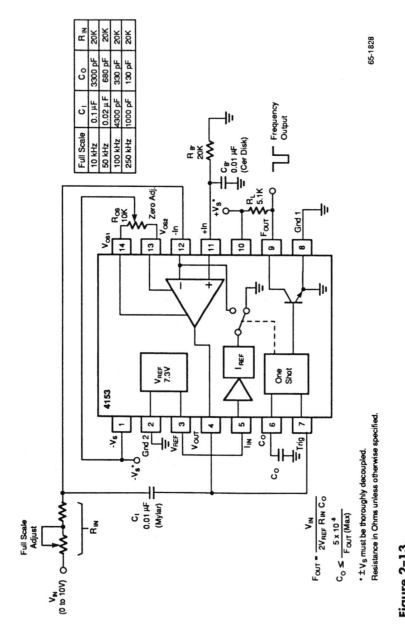

Full Scale	C_I	C_O	R_{IN}
10 kHz	0.1 µF	3300 pF	20K
50 kHz	0.02 µF	680 pF	20K
100 kHz	4300 pF	330 pF	20K
250 kHz	1000 pF	130 pF	20K

65-1828

$$F_{OUT} = \frac{V_{IN}}{2 V_{REF} R_{IN} C_O}$$

$$C_O \leq \frac{5 \times 10^{-5}}{F_{OUT} \text{ (Max)}}$$

* ±V_S must be thoroughly decoupled.
Resistance in Ohms unless otherwise specified.

Figure 2-13.
VFC with offset and gain adjustments (*Raytheon Semiconductor Data Book*, 1994, p. 3-816)

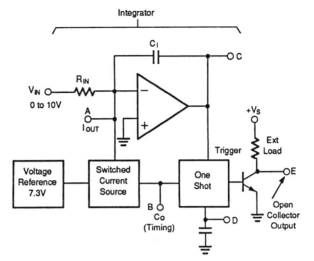

Figure 2–14.
4153 connected for VFC operation (*Raytheon Semi-conductor Data Book,* 1994, p. 3-817)

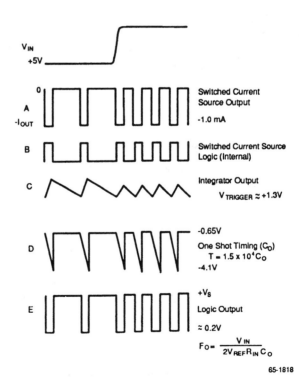

Figure 2–15.
4153 VFC timing wave-forms (*Raytheon Semiconductor Data Book,* 1994, p. 3-817)

2.2.2 Calibrating the VFC

The circuit of Fig. 2–13 is calibrated as follows. First, apply a measured full-scale input voltage and adjust R_{IN} for the desired scale factor. Then apply a small (but precise) input voltage (typically 10 mV) and adjust the op-amp offset until the output frequency equals the input multiplied by the scale factor. For example, assume that the input voltage range is 0 V to 10 V, and the desired full-scale output frequency is 250 kHz. Using the values for input and output capacitances shown in Fig. 2–13 (1000 pF for input, and 130 pF for output), adjust R_{IN} for 250-kHz output (R_{IN} should be about 20 k). Then reduce the input voltage to 10 mV, and adjust the offset for a 250-Hz output. Recheck calibration at the full-scale input of 10 V.

For applications in which precision is critical, the manufacturer recommends that trimpots (for R_{IN} and offset) be replaced by metal-film resistors soldered in parallel. As we all know, even the best of trimpots have bad temperature coefficients (TCs or tempcos), and are easily taken out of adjustment by mechanical shock.

2.2.3 Simplified Design Considerations for the VFC

The IC achieves its speed, accuracy, and temperature performance by incorporating high-speed ECL logic. However, the open-collector pull-up resistor (point E in Fig. 2–14) can be connected to a different supply (such as 5 V for TTL logic) as long as the voltage does not exceed the value of $+V_S$ applied to pin 10 (Figs. 2–11, 2–12, and 2–13). As shown in Fig. 2–15, the output at point E consists of negative-going pulses with a pulse width equal to the one-shot time (point D).

Load current should be kept below 10 mA to minimize strain on the IC. Also, pins 2 and 8 must be grounded in all applications, even if the open-collector transistor is not used.

2.2.4 Operating Principles (FVC)

Figure 2–16 shows the 4153 connected as a precision FVC. Figure 2–17 shows the corresponding waveforms and timing. This circuit converts the input frequency to a proportional voltage by integrating the switched current-source output. As the input frequency increases, the number of output pulses taken from the inverter increases, thus increasing the average output voltage. There is some ripple at the output, with ripple amplitude dependent on the integrator time constant. The output can be further filtered, but this reduces the response time. In general, a second-order filter will decrease ripple and improve response time, as discussed in Section 2.4.

For the circuit of Fig. 2–16 to operate properly, the input waveform must meet three conditions. First, the input must have sufficient amplitude and offset to swing above and below the 1.3-V trigger threshold. (The values of R_A and R_B must be selected to provide the correct offset, for a given value of $+V_S$.)

Second, the input must be a fast-slewing waveform with a quick rise time. This can be done with the comparator (to square up the input) and with AC coupling (capacitor C_A).

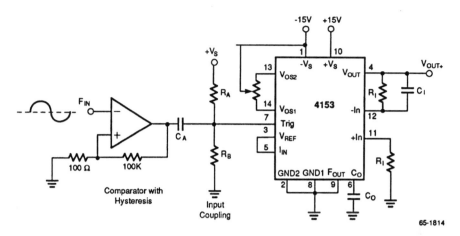

Figure 2-16. 4153 connected for precision FVC operation (*Raytheon Semiconductor Data Book,* 1994, p. 3-819)

Third, the input pulse width must not exceed the one-shot time (to avoid triggering the one-shot). Capacitive coupling between the trigger input (pin 7) and the timing-capacitor input (pin 6) might occur if the input waveform is a square wave, or if the input has a short time period. This can cause excessive nonlinearity because of changes in the one-shot timing waveform (as shown in Fig. 2–17). The problem can be avoided by keeping the value of C_O small. This will keep the timing period less than the input-waveform period. See Fig. 2–12 for some typical component values.

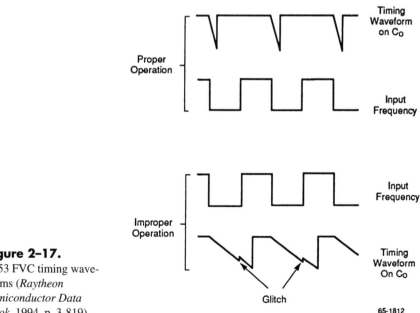

Figure 2-17.
4153 FVC timing waveforms (*Raytheon Semiconductor Data Book,* 1994, p. 3-819)

2.3 Comparison of 4151/4152 to 4153

Figure 2–18 shows the timing waveforms for the 4151/4152. Compare these to the waveforms of Fig. 2–15. Note that the input and output signal polarities are reversed. Also note that the 4153 is intended to be connected in the precision configuration, and thus contains an internal op amp. The 4153 current output is made negative so that the input voltage will always be positive. This function meets more user needs than the requirement of a negative input, and allows the use of high-performance NPN transistors in the switched current-source.

Because of the internal op amp, the 4153 requires fewer external components in the precision configuration. Of course, the 4152 can also be used in the precision circuit, but with an external op amp, as shown in Figs. 2–1, 2–2, and 2–7. The internal op amp of the 4153 provides improved specifications for temperature drift and linearity, especially at frequencies over 10 kHz (in addition to the obvious advantage of convenience).

The graphs in Figs. 2–19 through 2–24 show the differences in performance of the 4151, 4152, and 4153. The curves for the 4151/4152 are derived with the IC in the current-sourced precision circuit of Fig. 2–1. The voltage-sourced circuit, shown in Fig. 2–2, is slightly different because pin 1 (output current) is connected to ground rather than to the summing node of the op amp. When compared to the current-sour-

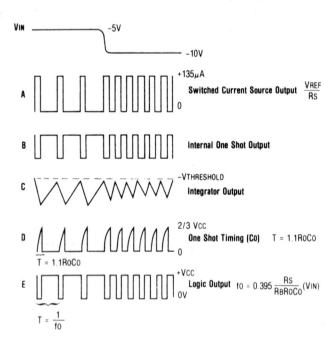

Figure 2–18. 4151/4152 timing waveforms (*Raytheon Semiconductor Data Book,* 1994, p. 7-7)

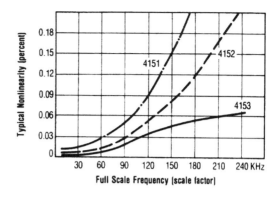

Figure 2–19.
Scale factor versus typical peak nonlinearity (*Raytheon Semiconductor Data Book,* 1994, p. 7-8)

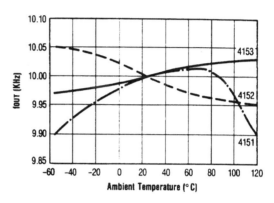

Figure 2–20.
10-kHz full-scale temperature drift (*Raytheon Semiconductor Data Book,* 1994, p. 7-8)

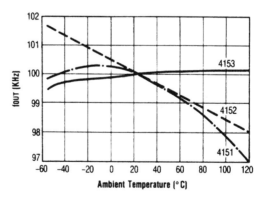

Figure 2–21.
100-kHz full-scale temperature drift (*Raytheon Semiconductor Data Book,* 1994, p. 7-8)

ced circuit of Fig. 2–1, the voltage-sourced circuit of Fig. 2–2 has better temperature performance, but at the expense of linearity.

Figure 2–19 shows how linearity is degraded with increasing full-scale frequency. Degradation occurs because of switching problems in the one-shot that affect the total amount of charge in each output-current pulse. Variations from the precise

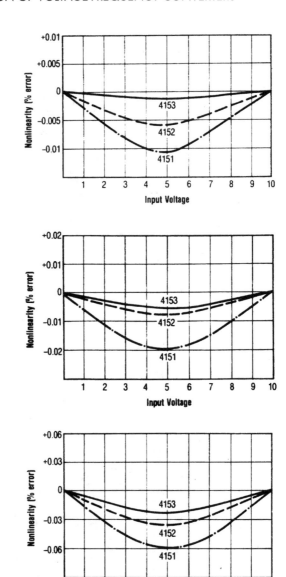

Figure 2–22.
10-kHz V-to-F non-
linearity (*Raytheon
Semiconductor Data
Book,* 1994, p. 7-8)

Figure 2–23.
50-kHz V-to-F non-
linearity (*Raytheon
Semiconductor Data
Book,* 1994, p. 7-8)

Figure 2–24.
100-kHz V-to-F non-
linearity (*Raytheon
Semiconductor Data
Book,*1994, p. 7-8)

charge cause deviations in the integrator output, affect the intervals between triggering, and shift the output frequency from the ideal. Because the 4153 uses high-speed
ECL, the high-frequency linearity is improved.

As shown in Figs. 2–20 and 2–21, temperature drift is affected by increasing
frequency. The 4153 outperforms the 4151/4152 with regard to temperature drift because the internal buried-zener reference of the 4153 provides the maximum tem-

perature stabilization. Of course, in any VFC/FVC IC, the reference, switched current-source, external resistor, and external capacitor temperature coefficients all contribute to temperature drift. So, as a simplified-design guideline, always use an external reference when temperature-drift specifications are fanatical! Use a temperature-stabilized external reference, such as an LM199. For best results, heat the 4153 with an external heater, measure temperature drift, then select resistors and capacitors for best temperature performance.

2.4 FVC Ripple Problems

As discussed, there is some ripple problem in any FVC circuit. The DC output contains a ripple component equal in frequency to the input pulses. This can be minimized with filtering, but with a decrease in response time. Consequently, there is always a tradeoff between response time and ripple. As a simplified-design guideline, use the largest amount of filtering that an acceptable response time will allow.

Figure 2–25 shows a FVC with a single-pole integrator at the switched current-source output. Figure 2–26 shows a second-order (double-pole) filter that will improve both response time and output ripple. The ratio of the time constants R1C1 and R2C2 determines the response to a step change in input frequency. The response is critically damped if R1C1 = 4(R2C2). Best results are obtained when R1C1 = R2C2, which provides a damping factor of one half.

Choose filter capacitors C1 and C2 (as well as the one-shot timing capacitor) for minimum ripple over the desired range of operation. As a simplified-design guideline for the 4153, peak ripple is less than 100 mV (10 Hz to 10 kHz) when R1 = 100 k and C1 = 0.1 µF.

Note that the values for the input capacitor and offset-bias resistors in Fig. 2–25 correspond to those shown in Fig. 2–12. As a simplified-design guideline, the time constant $C_A(R_A//R_B)$ should be kept less than 15 µS in most applications.

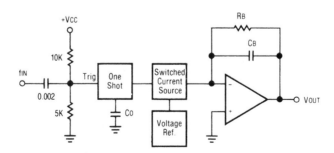

Figure 2–25. FVC with single-pole integrator (*Raytheon Semiconductor Data Book*, 1994, p. 7-9)

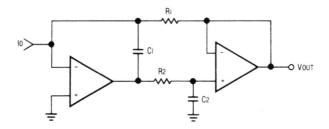

Figure 2-26. Second-order (double-pole) active filter (*Raytheon Semiconductor Data Book,* 1994, p. 7-9)

2.5 Integrators and A/D Converters

An integrator is essentially an A/D converter with a long sampling period. This concept makes it possible to produce an accurate analog integrator with a VFC connected to a counter and time base, as shown in Fig. 2–27. The signal to be integrated is converted to a frequency and counted over a known sample period. This provides a total count equal to the time integral of the signal, as shown by the equation in Fig. 2–27. Such an arrangement eliminates the need for expensive, low-drift, low-leakage op amps and capacitors found in analog integrators. This is especially true when the analog integrator must operate over a wide dynamic range, or integrate over extended periods.

Using the basic technique illustrated in Fig. 2–27, the signal to be integrated is converted to a frequency and then counted over a known sample period. This provides a total count equal to the time integral of the signal, as shown by the equation in Fig. 2–27. In addition to the ability to integrate over long periods of time, the VFC A/D-converter scheme shown in Fig. 2–27 has the following advantages:

- VFC A/Ds have high monotanicity because of their inherent linearity.
- The resolution and conversion speed are totally controlled by the system designer.

$$\int V_T = K \int F dt = K \int dN_T \qquad \frac{dt}{dt} = KN$$

where N = Total count
and K = VFC Scaling Constant

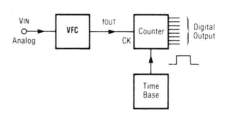

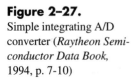

Figure 2-27.
Simple integrating A/D converter (*Raytheon Semiconductor Data Book,* 1994, p. 7-10)

- High noise rejection when integrating with a sample period longer than the noise period.
- Wide dynamic-input range because of inherent VFC linearity (10,000 to 1).
- The output can be directly interfaced to a digital information-processing system, used for a seven-segment display or converted back to analog with a low-cost D/A converter.
- The process can be interrupted without affecting the integrated value (no droop because of RC time constants).
- The digital counter can be preset to any desired value before integrating up (or down)

2.5.1 Using VFCs in Digital Voltmeter Applications

A digital voltmeter (DVM) is a classic A/D-converter application. (The voltage is converted to a digital value and then read out on a digital display.) One problem common to such circuits is that integration is good over long periods of time, but not during short periods. VFCs can be scaled to give many pulses in this short period, thus producing a high binary count to increase resolution. This is discussed further in Section 2.6.

2.5.2 Long-Term Integrator Design Example

Figure 2–28 shows the basic connections for a system that integrates the amount of sunlight on a photo cell over a 12-hour period. The maximum cell output is 5 V, and the visual display must be accurate to three digits.

Because it is easy to convert a binary count to decimal display with a binary-to-7-segment counter-driver, such as an ICM 7225, select a VFC scale factor that provides a count divisible by 10, after the 12-hour sampling period. There are 43,200 seconds in 12 hours, so 43,200 is the minimum binary count per volt. By using a maximum binary count of 5 million for 5-V input, a one-million binary count will equal one volt over 12 hours. One million pulses per 43,200 seconds equals a scale factor of 23.15 Hz per volt.

The pre-counter shown in Fig. 2–28 is used because the counter-driver typically has insufficient binary-counting stages. By moving the decimal six places, the display reads directly in volts. The time base can be made from 555 timer, as shown.

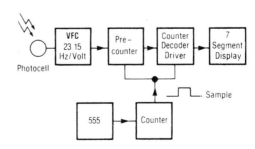

Figure 2–28.
Sun-power integrator
(*Raytheon Semiconductor
Data Book,* 1994, p. 7-11)

The timer must operate at a low frequency, and can be divided further by feeding the output to the clock input of a counter. This gives extremely long sample periods from the MSB output of the counter.

2.6 Low-Level A/D Converters

Figure 2–29 shows the basic connections for a low-level A/D converter using a VFC. This circuit avoids the small-signal resolution problem common to VFC A/D converters. Typically, the resolution of small signals is not as good as for large signals. This is because small voltages give a total count much less than the capacity of the counter, thus reducing accuracy even though the VFC output is highly accurate. The conventional solution to this problem is to add a preamp for small signals. However, this can cause errors, and must be switched out to measure large signals.

The system shown in Fig. 2–29 overcomes these problems by measuring the period (rather than the frequency) of the VFC output when small signals are to be processed. The system operates by counting a high-frequency reference oscillator during the interval (or period) between VFC pulses. The leading edge of the first VFC pulse toggles the flip-flop high, enabling the counter to begin counting the 1-MHz oscillator signals. The leading edge of the second VFC pulse toggles the flip-flop again, disabling the counter.

In the system of Fig. 2–29, the counter holds a binary count inversely proportional to the voltage input. Thus, the smaller the signal, the greater the resolution. Note that the input must be large enough to provide at least two output pulses so that the interval between pulses can be measured. The input can have a small DC bias applied, placing the input on the threshold of producing an output. This makes it possible to provide resolution in microvolts. The technique shown in Fig. 2–29 is very useful for measuring low-level transducer outputs (such as thermocouple outputs).

One problem with the circuit of Fig. 2–29 is that large signals applied to the VFC cause a small period, and thus lower resolution, because fewer oscillator pulses are counted. Figure 2–30 shows how the conventional VFC A/D can be combined with the system of Fig. 2–29. The circuit of Fig. 2–30 measures both the period and the frequency of the VFC output.

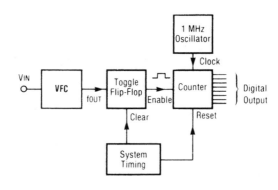

Figure 2–29.
Low-level A/D converter
(*Raytheon Semiconductor
Data Book,* 1994, p. 7-11)

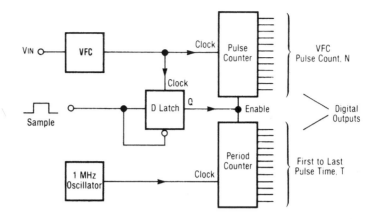

Figure 2–30. Measuring both period and frequency of VFC output (*Raytheon Semiconductor Data Book,* 1994, p. 7-11)

2.7 Integration Trend Indicator

Figure 2–31 shows a system that indicates integration at some intermediate time before completion of the sampling period (thus indicating a trend). (The conventional technique shown in Fig. 2–27 indicates the integration only after the sampling period is complete.) The system of Fig. 2–31 is still a long-term integrator but has the advantage of continuously updating the digital output with a short sampling period. The differential amplifier at the input subtracts the present V_{OUT} from the input. The difference signal is put through an absolute-value circuit and a scaling pot, which applies a positive voltage proportional to the magnitude of the differential input to the VFC.

During the sampling period, the up/down counter counts the VFC output frequency. The counter counts up if the present output is greater than the input, and vice versa. At the end of the sampling period, the count is latched into the input of the D/A converter, which now supplies the new output. The output of the latches can be used as the digital integrated output.

The time derivative of the integration (the degree of smoothing) is determined by the scaling of the inputs to the differential amplifier and by the scaling of the VFC input. The change of the output between samplings is proportional to the difference between the present output and the new input. The two counters are controlled by a D-type positive-edge-triggered flip-flop. The lower counter stores a count proportional to the time between the first pulse and the last pulse in the sampling period. The upper counter is clocked by the VFC output pulses and stores the total number of VFC pulses in the sample period. Deriving the period between pulses, and thus the input voltage, is simple arithmetic because the total time for a number of pulses is known.

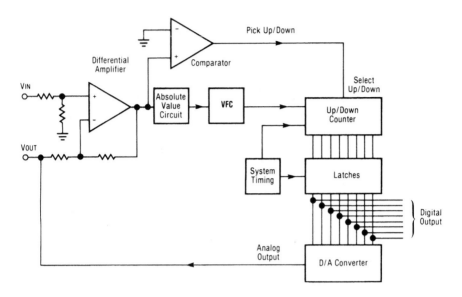

Figure 2–31. Integration trend indicator (*Raytheon Semiconductor Data Book*, 1994, p. 7-12)

2.8 Ratio Circuits

Figure 2–32 shows a circuit that provides an output proportional to the ratio of the two input signals. Such circuits can be used in power measurement and in determining transfer function (input to output, power-in to power-out, etc.).

The system of Fig. 2–32 uses two VFCs and some digital processing to produce a binary number proportional to the ratio of two input signals. This digital output can be used directly or converted back to a voltage with a D/A converter. $V_{IN}2$ is converted to a square wave by the flip-flop. This square wave (divided by 2) is used

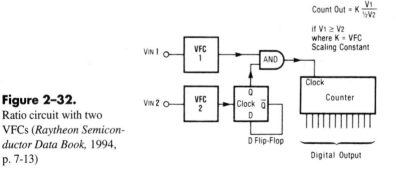

Figure 2–32.
Ratio circuit with two VFCs (*Raytheon Semiconductor Data Book*, 1994, p. 7-13)

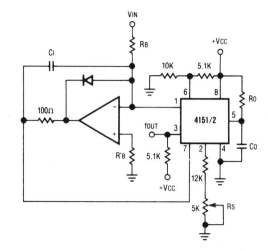

Figure 2–33.
Ratio circuit with one
VFC and precision resis-
tance (*Raytheon Semicon-
ductor Data Book,* 1994,
p. 7-13)

to gate the output of VFC 1. Thus, the number of pulses from output one (per pulse of output two) is counted. Note that the frequency from VFC 1, and thus the input to VFC 1, must be more than or equal to that of VFC 2. Also, system timing must be tailored to individual applications.

The circuit of Fig. 2–33 uses resistance ratios instead of two VFCs. The circuit produces an output frequency that is directly dependent on the ratio of R_S/R_B when all other factors are held constant. The equation for the output frequency of the 4152 is:

$$\text{Output frequency } (f_O) = (0.486/R_O C_O) \, (R_S/R_B) \, V_{IN}$$

The resistor to be trimmed is connected to the R_S position. The reference resistor is connected to the R_B position. When V_{IN} is adjusted to –9.76 V, the output frequency will be 10 kHz with a one-to-one resistance ratio. See Figs. 2–1 through 2–4 for typical component values. Note that the 4153 cannot be used in the circuit of Fig. 2–33 because the R_S resistor is internal and cannot be adjusted. The output of the Fig. 2–33 circuit can be compared to a reference frequency, or read on a frequency counter.

The circuit of Fig. 2–33 can also be used to measure capacitance ratios. In this case, R_S, R_B, R_O, and V_{IN} are held constant. The output frequency is then proportional to $1/C_O$, and the period of the output frequency is proportional to C_O. If the need is to compare a capacitor with a reference capacitor, the circuit can be adjusted for a given output frequency with the reference C_O in place. Then C_O is replaced with the capacitor to be measured and the difference in output frequency is measured. The difference (or beat) frequency can be monitored as an audio signal, as well as read with a counter.

2.9 Microprocessor-Controlled DVM

Figures 2–34, 2–35, and 2–36 show the block diagram, schematic, and program listing for a microprocessor-controlled DVM. The circuit uses the 4153 and an Ohio Scientific CIP microcomputer. The 6502 microprocessor in the CIP controls the counters, and puts the data in memory for display on a CRT.

Using the values shown, the 4153 is connected for 1-kHz full-scale operation. The output at pin 9 is used to clock two four-bit counters which, in turn, are buffered into the microprocessor data bus. Control of the buffers is taken from the microprocessor address bus through hard-wired logic and an eight-bit latch.

The program (Fig. 2–36) first clears the display, takes 10 sample readings and averages them, then displays the result. The following is a step-by-step summary of the program listing:

1. The first four lines initialize the program, clear the output variables, and erase the display. Line 50 sets up the loop for the averaging routine.
2. Lines 60 through 100 take the actual measurement. (The hard-wired logic addresses were selected by checking the memory map for unused locations.) Line 60 addresses location F7FF, which latches in the binary number 0 from the data bus. The latch output resets the two binary counters to zero, and keeps the counters from counting until the next program line.
3. Line 70 again addresses the latch, and the data-bus word enables the count to begin. The next two lines are a loop to set the sample period, which ends when the latch is again addressed. The counters now contain a binary number proportional to the VFC output frequency. Note that there are 6 unused latch outputs that can be used for other purposes.

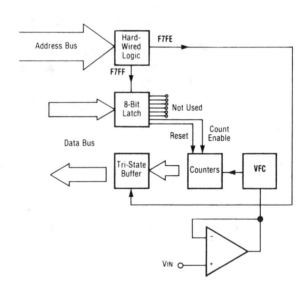

Figure 2–34.
Microprocessor-controlled DVM (*Raytheon Semiconductor Data Book,* 1994, p. 7-14)

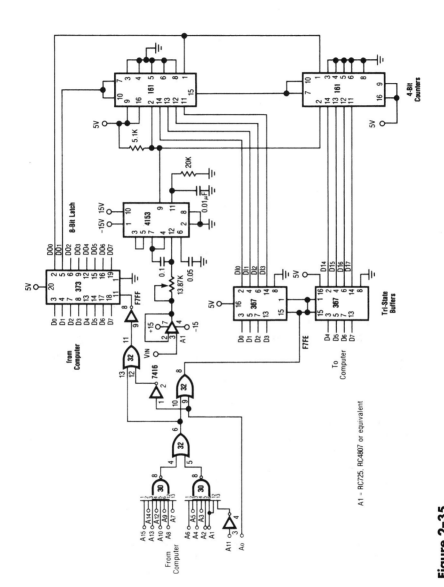

Figure 2-35.
Microprocessor-controlled DVM schematic (*Raytheon Semiconductor Data Book*, 1994, p. 7-15)

```
REM = Remark
10  X=236
20  S=10
30  FOR Y=1 TO 5:PRINT:NEXT
40  A1=0
50  FOR V=1 TO 5
60  POKE 63487,0:REM Clear Counters
70  POKE 63487,255:REM Start to Count
80  FOR T=0 TO X:NEXT
90  POKE 63487,253:REM Stop Count
100 A=PEEK(63486)
110 A=10.004*A/253
120 PRINT A
130 A1=A1+A
140 NEXT V
150 A1=A1/S
160 A1=.01*INT(100*A1)
170 PRINT"READING";A1;"VOLTS"
180 PRINT
190 GOTO 30
```

Figure 2–36.
Microprocessor-control-
led DVM program listing
(*Raytheon Semiconductor
Data Book,* 1994, p. 7-16)

4. The program now addresses F7FF, which selects the tri-state buffers and loads the counter outputs onto the data bus. The measured count is converted to an equivalent voltage reading by finding the ratio of that count to the full-scale count, and multiplying by what has been defined as the full-scale input voltage. This is shown by:

$$V_{IN} = N/(N_{FS}) (V_{FS})$$

where N = present count, N_{FS} = full-scale count, and V_{FS} = V_{IN} for full-scale count. In the circuit of Fig. 2–35, 10.004 V is applied to the VFC, and the sampling period is adjusted until the counters count without going into overflow. (In this case, the count is 253 because of non-ideal timing increments.)

As an example, when 3 V is applied to the VFC (with an approximate 10-V full scale), the counter will count to 0.3 of full scale, or 76. Accuracy to within three digits is obtained by dividing 76 by 253, which equals 0.30039. Multiplying the result by 10.004 yields 3.005 V. When this is combined with the program averaging techniques, there is a consistent accuracy of three digits.

Line 110 provides this computation, and lines 120 through 150 are part of the 10 reading averaging technique. Line 160 cuts the unneeded digits (because the accuracy is limited by the 8-bit data capacity). The last lines display the final averaged result, and loop back to the beginning of the program. The value of S set in line 20 determines the number of samples averaged. Line 120 prints the individual voltage readings, and can be deleted after debugging is complete.

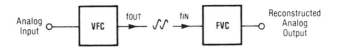

Figure 2–37. VFC to FVC transmission (*Raytheon Semiconductor Data Book,* 1994, p. 7-17)

The system described here was constructed (by Raytheon) using two separate PC boards: one for the interface and the other for the 4153 and counters. All ICs in the interface can be used with any microprocessor with a compatible bus, provided that the two address codes are in unused locations.

The voltage follower A_1 shown in Fig. 2–35 is used to minimize circuit loading, and can be omitted in some cases. Also, the 1-kHz full-scale operation used in this example can be altered if desired. As always, use low TC components, well-bypassed power supplies, and good grounds to minimize temperature drift and nonlinarity.

2.10 Telemetry and Data Transmission

VFCs and FVCs are well suited to telemetry, data transmission, and other remote-monitoring applications. As shown in Fig. 2–37, a VFC can be used to convert analog signals directly to serial form. The serial data bits can then be transmitted to a FVC for reconstruction of the analog signal, or can be converted directly to digital by a counter and latch. The open-collector output of a VFC can be used directly for twisted-pair transmission lines up to several hundred feet. For distances of several thousand feet, a line driver and receiver can be used. Longer distances require radio or telephone FSK (frequency shift keying) transmission.

There are many problems (and solutions) when voltage-frequency converters are used in telemetry and data transmission. The remaining paragraphs of this section describe the major problems and some practical solutions.

2.10.1 Transmitter-Receiver Isolation

One major problem is that some applications require electrical isolation between the transmitter and receiver. This is especially true when the transmitter and receiver operate at entirely different voltage levels, and there is a high-voltage difference between the two. The classic solution is to use an isolation amplifier and an A/D converter. However, for high accuracy (12 bits or more), the cost becomes prohibitive when compared to the voltage-frequency converter solution.

Figure 2–38 shows the basic circuit for isolating a transmitter and receiver. The VFC is powered by a floating power supply and transmits a pulse train to the medium. The medium is usually an optoisolator (LED and photo transistor in one package), but could be a transformer or even a speaker and microphone. The medium can also serve as the transmission line (optical fiber, infrared, or radio). Pulses from the medium can be reconverted with an FVC or used for direct digital output.

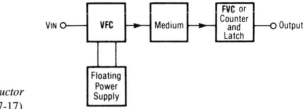

Figure 2–38.
VFC/FVC isolation
(*Raytheon Semiconductor
Data Book*, 1994, p. 7-17)

2.10.2 Optoisolators and Fiberoptic Transmission

Figure 2–39 shows a basic optoisolator circuit for a VFC. This circuit provides isolation between the transmitter and the rest of the system. Such circuits are often used to protect microcomputers from transient voltages and grounding problems, or where a transducer has a high voltage.

The optoisolator is pulsed directly by the VFC at the relatively low level of 10 mA, conserving power. One stage of gain is added for the logic output. The light emitted by the LED provides base current to turn on the phototransistor. This turns off the output transistor and produces a logic high at the output. The basic circuit has a wide bandwidth and will drive at least six standard TTL loads.

Figure 2–40 shows a complete fiberoptic transmission circuit using a VFC as the transmitter. Transistor Q_1 is added at the transmission end to provide increased current capability, and to invert the negative output pulses from the VFC. Q_1 must be selected for a current capability equal to the requirements of the transmitter.

Pulses of light are sent through the fiber to be received by the photodiode. The light received by the photodiode is converted into a small current. This current is amplified by A_1 and applied to comparator A_2, which has an adjustable reference to set the threshold for the logic output. The circuit of Fig. 2–40 has a bandwidth sufficient to handle 20-kHz signals.

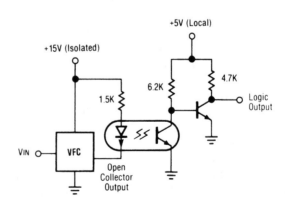

Figure 2–39.
VFC optoisolator
(*Raytheon Semiconductor
Data Book*, 1994, p. 7-17)

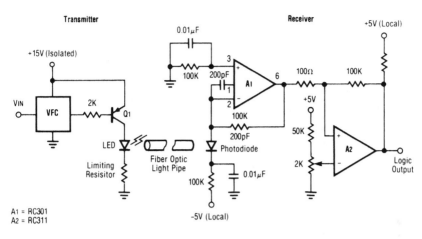

Figure 2–40. Fiberoptic transmission with VFC (*Raytheon Semiconductor Data Book,* 1994, p. 7-18)

2.10.3 Frequency-Shift-Keying Modulators

A common requirement for telemetry and other forms of data communications is the transmission and decoding of binary data as two or more discrete frequencies. This frequency-shift-keying (FSK) function can be implemented with a VFC, two transistors, and a flip-flop.

Figure 2–41 shows an FSK modulator using the 4153 VFC. Figure 2–42 shows the component values for operation over public telephone lines using the 300-baud Bell 103 Standard.

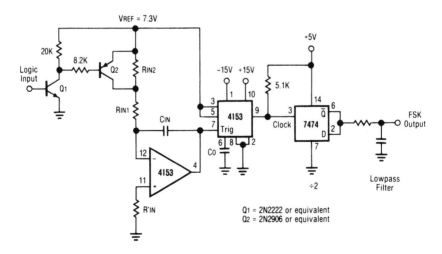

Figure 2–41. 4153 FSK modulator (*Raytheon Semiconductor Data Book,* 1994, p. 7-18)

Figure 2–42.

Component values for
4153 FSK modulator
(*Raytheon Semiconductor
Data Book,* 1994, p. 7-19)

Mark = 1070Hz	Mark = 2025Hz
Space = 1270Hz	Space = 2225Hz
RIN1 = 28.3K	RIN1 = 16.1K
RIN2 = 6.0K	RIN2 = 1.6K
Co = 0.0068μF	Co = 0.0068μF
CIN = 0.1μF	CIN = 0.1μF

When the logic input goes high, Q_1 is turned on, as is Q_2. The input current is equal to:

$$V_{REF} - 0.2 \text{ V/R}_{IN1}$$

The one-shot frequency increases to balance this current.

If the logic input goes low, Q_2 turns off. The integrator input current is:

$$V_{REF}/R_{IN1} + R_{IN2}$$

so the one-shot frequency (and thus the logic-output frequency) decreases to balance the loop.

The output pulse is applied to the toggle, which divides by two, and provides an output square wave at one-half the VFC frequency. The toggle output is applied to the telephone line through a lowpass filter (to round the sharp output pulses from the toggle).

Figure 2–43 shows an FSK modulator using the 4151/4152 VFC. Figure 2–44 shows the component values for operation over public telephone lines using the 300-baud Bell 103 Standard.

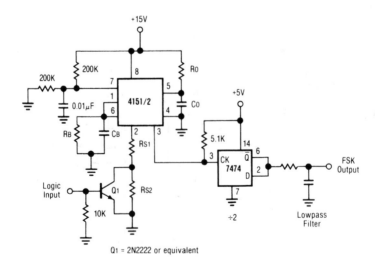

Figure 2–43. 4151/4152 FSK modulator (*Raytheon Semiconductor Data Book,* 1994, p. 7-9)

Figure 2–44.

Component values for 4151/4152 FSK modulator (*Raytheon Semiconductor Data Book,* 1994, p. 7-19)

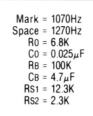

| Mark = 1070Hz |
| Space = 1270Hz |
| R_O = 6.8K |
| C_O = 0.025μF |
| R_B = 100K |
| C_B = 4.7μF |
| R_{S1} = 12.3K |
| R_{S2} = 2.3K |

| Mark = 2025Hz |
| Space = 2225Hz |
| R_O = 6.8K |
| C_O = 0.025μF |
| R_B = 100K |
| C_B = 4.7μF |
| R_{S1} = 23.2K |
| R_{S2} = 2.3K |

When the logic input goes high, Q1 turns on and shorts out R_{S2}, thus changing the scale factor. Overall frequency trim can be changed by trimming R_B. Adjust R_{S2} for relative frequency trim. The output frequency (F_O) is equal to:

$$R_S V_{CC}/2.53 R_O C_O R_B$$

It may be necessary to use an inverter at the input to keep logic polarity correct for a particular system. With the circuit of Fig. 2–43, the output frequency decreases when a logic high is applied at the input. Also, components for the lowpass filter at the output must be selected with the bandwidth and response time limitations of the transmission medium in mind.

2.11 Digitally Controlled Signal Generator

Figure 2–45 shows how a VFC can be connected with a D/A converter to produce a digitally-controlled signal or clock generator. The binary input code is converted to a voltage by the D/A converter, and then applied to the VRC. In turn, the VFC output is converted to a square wave (and divided by two) by the toggle circuit.

Figure 2–46 shows the details of a digitally-programmable clock generator using a 4153. The circuit provides a 100-Hz to 9.9-kHz square wave output programmed by thumbwheel switches, or by an 8-bit BCD input word. The output changes in 100-Hz steps, one step for each least-significant-digit change. The DAC-20 is a 2-digit BCD D/A converter which provides a 0 to 1-mA current output. The reference voltage for the D/A is provided by the 4153.

Amplifier A_1 converts the D/A current output to a 0 to 9.9-V signal applied to the 4153. Scale factor for the VFC is 2 kHz/volt, which is then divided by two by the

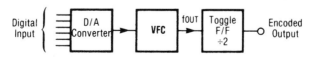

Figure 2–45. Basic digitally-controlled frequency generator (*Raytheon Semiconductor Data Book,* 1994, p.7-20)

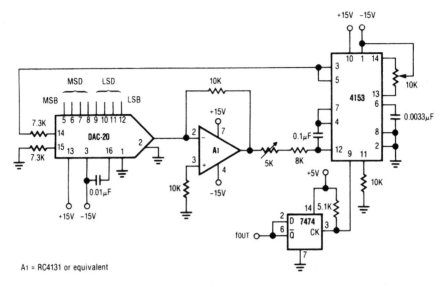

Figure 2–46. Digitally-programmable clock generator (*Raytheon Semiconductor Data Book,* 1994, p. 7-20)

7474 latch (connected as a toggle). This produces a square-wave output of 100 Hz to 9.9 kHz. The VFC scale factor can be adjusted to give the desired range.

To calibrate the circuit, set the D/A inputs to maximum (1001 1001), and trim the 5-k pot until the output is 9.9 kHz. Then set the D/A inputs to minimum (0000 0001) and adjust the offset (10-k pot) until the output reads 100 Hz. The 4153 linearity characteristics ensure that all other settings are in calibration (maximum to minimum). The circuit of Fig. 2–46 can also be used with other types of D/A converters, as well as different frequency scales and output signal conditioning.

2.12 Bipolar Telemetry

Figure 2–47 shows how a VFC can be connected with various components to produce a bipolar telemetry transceiver. Such a circuit provides for serial transmission of bipolar analog signals over a twisted pair (or any other form of transmission medium). This circuit overcomes the usual limitations of bipolar serial transmission by encoding the pulse output.

Negative input voltages are transmitted as a negative-going pulse train, and positive voltages as a positive-going train. This information is decoded into a frequency output proportional to the magnitude of the input, and a sign output that depends on the polarity of the input. The absolute-value circuit applies a voltage proportional to the magnitude of the input to the VFC. The frequency output of the VFC is encoded with the polarity output of the comparator, giving a coded signal that is transmitted to the decoder.

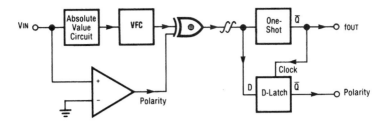

Figure 2–47. Bipolar telemetry transceiver (*Raytheon Semiconductor Data Book,* 1994, p. 7-21)

Figures 2–48, 2–49, and 2–50 show the bipolar-telemetry receiver, transmitter, and waveforms, respectively. The absolute-value circuit in the transmitter provides a positive voltage equal to the magnitude of the input voltage. The buffered output is applied to the VFC. In turn, the VFC generates 10-μs wide negative-going pulses with a scale factor of 1 kHz/volt. Full-scale input is therefore 20 Vp-p (that is, peak-to-peak).

The transistor provides a logic output that depends on the polarity of V_{IN}. (Section 2.17 describes additional information on absolute-value circuits.) The polarity signal is used to encode the frequency output of the VFC in a TTL exclusive-OR gate. Positive-going pulses are generated from positive inputs, and vice versa.

The coded pulse information is transmitted and applied to the D input of the 7474 latch, and to the two trigger inputs of the dual-edge retriggerable one-shot (8853). For the components shown, the one-shot generates a 25-μs negative-going pulse for each 10-μs input pulse. The one-shot triggers for negative-going or positive-going pulses on the last edge encountered. This produces a 25-μs pulse train equal in frequency to the data, as shown in Fig. 2–50.

The rising edge of the pulses clocks the D-latch, and the information on the D-latch input is latched in. If the pulses are negative-going, the clock will rise when the D-input is high, latching the Q-output low. If the pulses are positive-going, the clock

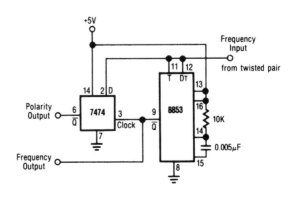

Figure 2–48.
Bipolar telemetry receiver (*Raytheon Semiconductor Data Book,* 1994, p. 7-21)

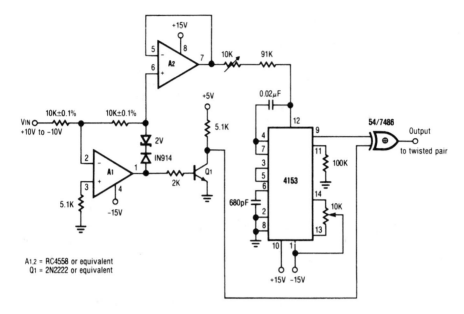

Figure 2–49. Bipolar telemetry transmitter (*Raytheon Semiconductor Data Book,* 1994, p. 7-21)

rises when the D-input is low, latching the Q-output high. The latch output then corresponds to the polarity of the data transmitted.

2.13 Two-Wire Transmitter for Telemetry Transducer

Figure 2–51 shows a complete two-wire transmission system using the 4151/4152. The same wires that carry the data signal also supply power to the VFC, eliminating batteries or isolated power supplies at the transducer. The circuit shown

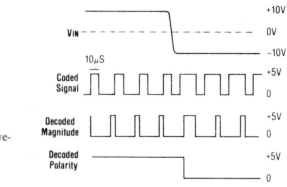

Figure 2–50.
Bipolar telemetry waveforms (*Raytheon Semiconductor Data Book,* 1994, p. 7-22)

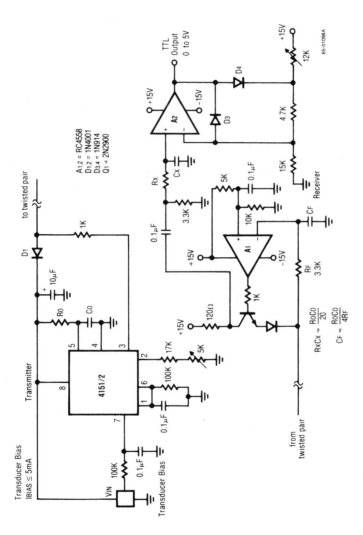

Figure 2-51.
Two-wire transmission system (*Raytheon Semiconductor Data Book*, 1994, p. 7-22)

uses the 4151/4152 in the single-supply configuration, delivering 1% accuracy. Greater accuracy can be obtained by adding a single-supply op amp for the precision mode (Section 2.1). The components shown in Fig. 2–51 provide for 0 to 10-kHz operation.

The output of the 4151/4152 pulls current pulses from the transmission line, which are detected and output as a TTL signal by the receiver. The blocking diode D_1 and 10-μF capacitor filter the supply to the VFC and absorb line reflections of the transmitted data. Bias voltage for the transducer can be provided the transmitter supply if the current requirement is less than 5 mA.

Transistor Q_1 acts as a series-pass circuit to regulate the transmitter supply, and functions as a common-base amplifier for the data pulses. The current pulses create a voltage drop across the 120-ohm resistor, which is capacitively coupled into a filter and comparator. In turn, the comparator is connected so that the output swings from 0 V to +5 V (TTL logic). $C_F R_F$ and $C_X R_X$ provide noise filtering to reject electromagnetic interference from motors and power lines. $C_F R_F$ controls the response time of the power regulator (A_1 and Q_1), and $C_X R_X$ provides noise filtering to reject electromagnetic interference from motors and power lines. $C_F R_F$ controls the response time of the power regulator (A_1 and Q_1), and $C_X R_X$ provides a lowpass filter for the output amplifier. D_2 protects Q_1 from transient voltages on the transmission line. Typical values for a 75-μs pulse width are: $C_F = 0.005$ μF, $R_X = 1$ k, $C_X = 0.001$ μF.

2.14 Tape Recording with a VFC

Figure 2–52 shows how a VFC can be used to record data (such as telemetry-transducer data) on magnetic tape. An application for this circuit might be the taking of data at a remote site with only battery-operated equipment available, or file storage for later reference. The data can be retrieved for use in digital format (the toggle output), or reconverted (by an FVC) to analog for chart-recorder display.

The VFC output (pulses at a frequency corresponding to the analog input) is applied to the toggle flip-flop. This results in a square wave of half the frequency from the VFC. The signal is applied to a lowpass filter to round the square wave, because sharp rise-times will cause the NAB compensation (found on most tape recorders) to overload the tape.

Both high and low frequency response limit the dynamic range of the input, so it might be necessary to add a DC offset to the transducer (or other analog input to the

Figure 2–52.

Basic VFC tape-recording interface (*Raytheon Semiconductor Data Book*, 1994, p. 7-23)

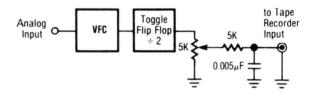

VFC) to get a frequency offset. Best results should be obtained when the recorded signal is in a 1- to 5-kHz range.

The signal can be retrieved from the tape by capacitive coupling of the output to a ground-referenced comparator. This will give a square-wave output that can be processed digitally, or applied to an FVC for analog reconstruction. Remember that the FVC must have twice the scale factor of the VFC for 1-to-1 reproduction (because the flip-flop divides by two).

Certain restrictions and signal-conditioning problems must be observed when using the circuit of Fig. 2–52. The frequency response of the tape deck (and not the circuit) limits the dynamic range of the analog data. Most recorders will not accept pulse trains of sharp square waves (because of NAB or other compensation schemes). Accuracy of the system is limited by motor-speed variations (wow and flutter) of the deck. Good battery-operated cassette recorders will introduce about 1 to 3% inaccuracy. If battery operation is necessary, make certain that the batteries are well charged. If not, tape speed can vary as the batteries discharge.

Many cassette decks have a 200–300 Hz AC component in the motor speed. This is a function of the motor poles and belt ratios. If higher accuracy is essential, use high-quality portable recorders with an alleged 0.5% or better accuracy.

2.15 Motor Speed Controls

There are two conventional approaches to providing control of motor speed with voltage-frequency converters. The system shown in Fig. 2–53 uses an FVC to provide an error signal that depends on variations in motor speed. Figure 2–54 shows how a VFC is used to provide open-loop switching control of motor speed. In addition to these two basic methods, three other motor-speed control techniques use voltage-frequency converters. These include: pulse-width-modulation (PWM) speed control, series-pass summing control, and phase-locked loop speed control, all of which are discussed in this section.

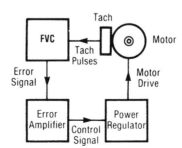

Figure 2–53.

Motor-speed control with FVC feedback (*Raytheon Semiconductor Data Book*, 1994, p. 7-24)

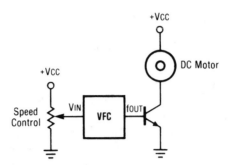

Figure 2–54.
Open-loop motor-speed control (*Raytheon Semiconductor Data Book,* 1994, p. 7-25)

2.15.1 Control with FVC Feedback

As shown in Fig. 2–53, tachometer pulses are converted into a voltage that depends on motor speed. This signal is amplified (with respect to some reference voltage) to provide a control signal that regulates the power supplied to the motor. If the motor slows down, the DC voltage from the FVC decreases, causing the control signal to increase. This makes the circuit deliver more power, causing the motor to speed up.

2.15.2 PWM Control with FVC Feedback

Figure 2–55 shows a more sophisticated version of the basic FVC-feedback system. Figure 2–56 shows the related waveforms. This circuit uses an optical tachometer and a comparator to provide pulse-width modulation (PWM) for switching-control of a DC motor. The circuit has better rejection of variations in load, power-supply voltage, and temperature than the basic system.

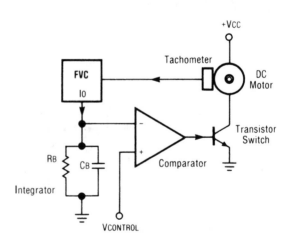

Figure 2–55.
Precision FVC motor-speed control (*Raytheon Semiconductor Data Book,* 1994, p. 7-24)

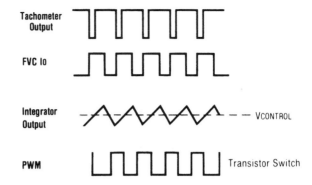

Figure 2–56.
Precision FVC motor-
speed control waveforms
(*Raytheon Semiconductor
Data Book*, 1994, p. 7-25)

The tachometer sends a pulse train to the FVC, which produces a ramp signal with a DC level that is proportional to motor speed. The ramp is compared to a control input to generate a PWM signal to drive the motor switch. The values of components depend on the tachometer, motor, and specific application. The following are some typical guidelines for component values.

For smooth operation, the tachometer frequency should be much greater (at least 10 times) than the motor RPM. This will ensure that the load friction/inertia product does not slow down the motor between tachometer pulses. If necessary, a frequency multiplier can be used to make the apparent ratio of tach frequency to motor revolutions higher. Of course, the tach pulses must be within the input-voltage constraints for the FVC circuit (as discussed in Section 2.17.3).

The FVC output-current pulses (with amplitude I_O and width T_P) are integrated by the parallel circuit of $R_B C_B$. The time constant of $R_B C_B$ is chosen such that the input to the comparator is a triangle wave. This triangle must have a peak-to-peak amplitude large enough to provide good pulse-width modulation, but also have a time constant long enough to provide an average voltage that is proportional to motor speed.

The approximate peak-to-peak voltage is given by:

$$V_{p-p} = V_{high} - V_{low} = I_O/C_B \; T_P \; 1 - (T_P/T)$$

where T_P = one-shot time (1.1 ROCO) and T = 1/tach frequency

The average voltage is given by:

$$C_{comp} \text{ (average)} = R_B I_O \; (T_P/T)$$

Typically, you must experiment with the R_B and C_B values to match the motor and tachometer characteristics. It is usually easier to start by finding the average voltage first.

The speed-control input ($V_{control}$) is compared to the filtered FVC output, which contains a triangular ripple component and a DC component proportional to motor speed. Variations in motor demands (or supply voltage VCC) or a change in the control input causes the comparator output to vary in duty cycle (classic PWM). If the motor speeds up, the power transistor is switched on for a shorter time, thus slowing the motor down.

Component values are chosen to provide an approximate 30 to 50% duty cycle under normal loading. The comparator must have high gain for positive switching, and must be able to drive the output transistor into saturation. For larger motors, add more output transistors (in parallel) to accommodate motor requirements.

2.15.3 Open-Loop Switching Control

The circuit of Fig. 2–54 uses a VFC to provide open-loop (no feedback) speed control of a DC motor. The frequency of the pulses provided by the VFC depend on the speed-control voltage. The pulses are used to switch a power transistor, providing pulsed DC power to drive the motor. This increases the motor efficiency, and allows smoother operation (at lower RPMs) than a steady DC supply can provide. However, this pulsed or AC system has a problem of heating the motor if it is used for extended periods.

2.15.4 Series-Pass Summing Control

The circuit of Fig. 2–57 combines the low-speed smoothness of Fig. 2–54 with the advantages of DC power (low heating). The summing-amplifier output contains an AC component produced by the VFC, and a DC component proportional to the speed-control voltage. This voltage, minus the base-emitter voltage of the pass transistor, appears across the motor. The inputs to the summing amplifier are scaled so that the pulse train is eliminated when the speed-control voltage reaches about 50% of the full-scale value. (The summing amplifier must be non-inverting.)

2.15.5 Phase-Locked-Loop Speed Control

Figure 2–58 shows another PWM scheme that has the advantages of increased gain at higher RPMs and a tach frequency that equals the VFC output frequency. Figures 2–59 and 2–60 show the PWM waveforms and phase-comparator relationships, respectively.

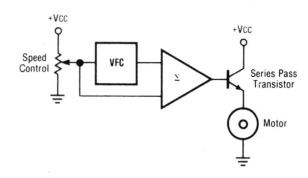

Figure 2–57.
Series-pass summing control (*Raytheon Semiconductor Data Book,* 1994, p. 7-26)

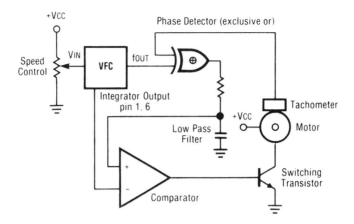

Figure 2–58. Basic PLL speed control (*Raytheon Semiconductor Data Book,* 1994, p. 7-26)

The VFC converts the speed-control voltage into a reference frequency with which to compare the tachometer frequency, and provides a triangle waveform used to pulse-width modulate the output. The VFC frequency is compared to the tachometer frequency in a digital exclusive-OR phase detector. For proper phase comparison, the VFC and tachometer outputs must both have 50% duty cycles when locked. The phase-detector output is fed through a low-pass filter to provide a varying-DC error signal at the comparator input. This comparator functions much like that of the comparator in Fig. 2–55 to provide PWM at the output. The low-pass filter must have a longer time constant than that of the VFC integrator to provide good pulse-width modulation.

Because the triangle waveform (Fig. 2–59) is equal in frequency to the VFC output, the VFC frequency is not only the same as the tach frequency, but also the drive-pulse rate. Therefore, to ensure smooth operation, the tach frequency must be greater than the motor RPM rate. As shown in Fig. 2–60, the phase-comparator inputs will vary in angle from 0° to 180°. At 0°, the low-pass filter output is 0 V. At 180°, the output is maximum. To provide a proper error signal, the PWM must be scaled so that the system will balance when the phase angle is 90°.

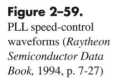

Figure 2–59.
PLL speed-control waveforms (*Raytheon Semiconductor Data Book,* 1994, p. 7-27)

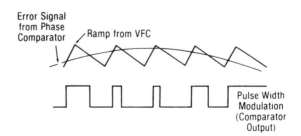

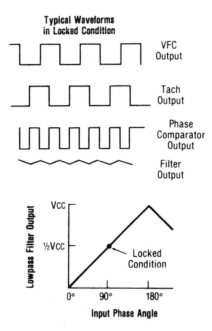

Figure 2–60.
Phase-comparator rela-
tionships (*Raytheon
Semiconductor Data
Book,* 1994, p. 7-27)

This type of phase comparator might lock onto harmonics of the tach input. The range of frequencies over which the comparator will lock depends primarily on the relationship of the low-pass filter and the time constants of the motor drive. As always, it is necessary to experiment with filter characteristics, tach-pulse to motor-RPM ratio, and scaling of the PWM inputs to get best results.

In the locked condition, if the motor sees a greater load and slows, the filter output voltage increases and provides more on-time to the output transistor. This speeds up the motor. The peak amplitude of the triangle waveform decreases as the VFC frequency increases.

For a 50-to-1 frequency change, the peak-to-peak triangle voltage is 3.8 times smaller. This reduction of input voltage to the comparator acts as an increase in loop gain at high frequencies. The 11.6-dB effective gain increases at the high end compensate for the gain reduction at high motor speeds (a condition inherent in all PWM DC motor-drives).

2.16 Miscellaneous Applications

As discussed throughout this book, there are many applications for VFCs and FVCs other than basic voltage-frequency conversion. This section describes three such applications.

2.16.1 Controlling a Chemical Reaction (Chloride Titrator)

Figure 2–61 shows how the precise charge-dispensing characteristics of a VFC are used to control a chemical reaction. Current pulses from the VFC generate Ag+ ions, which precipitate Cl⁻ ions from the electrolyte solution. If there are Cl⁻ ions in the solution, the A$_g$ electrode is polarized positive. This switches the ground-referenced comparator, which fires the one-shot, switches the logic output, and dumps a charge from the A$_g$ electrode into the solution.

The charge causes Cl⁻ ions to be absorbed by the A$_g$ electrode, so the one-shot keeps firing until all Cl⁻ ions are gone. When that happens, the A$_g$ electrode is polarized negative, below the comparator threshold, and the one-shot stops firing. The number of charges dispensed, which equals the number of logic-output pulses, is a measure of how many Cl⁻ ions were in the solution.

To convert the circuit of Fig. 2–61 into a practical measurement system, a titration of a standard solution of Cl⁻ is necessary for calibration. This requires a knowledge of chemistry far beyond that of the author's limited range!

2.16.2 Stabilizing a VCO

Figure 2–62 shows how an FVC can be added to the feedback loop of a VCO (voltage controlled oscillator) to improve performance. (Low-cost VCOs usually do not have good temperature characteristics.) The circuit of Fig. 2–62 can provide a linearity of 0.01% and a low temperature coefficient, without using heating elements or expensive digital synthesis.

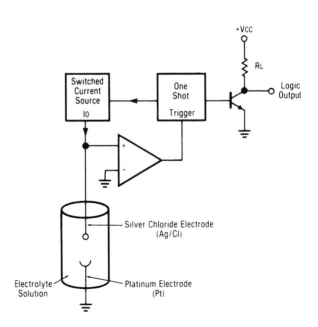

Figure 2–61.

Basic chloride titrator (*Raytheon Semiconductor Data Book,* 1994, p. 7-27)

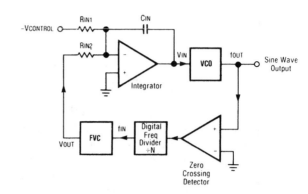

Figure 2–62.
Stabilizing a VCO with
an FVC (*Raytheon Semi-
conductor Data Book,*
1994, p. 7-28)

The VCO output is a high-frequency sine wave. This is squared in a comparator and the frequency is digitally divided down to a frequency usable by the FVC.
Some signal conditioning might be necessary at the FVC input (refer to Section
2.17.3).

The FVC provides an error signal that is applied to a summing integrator. The
FVC output is balanced with the V-control signal to produce the voltage-control input
to the VCO. If the VCO drifts, the FVC output voltage changes, causing the integrator output to produce an error signal for the VCO. Low temperature coefficient components should be used in the FVC and integrator (such as a low-drift op amp and
polystyrene capacitors).

2.16.3 Staircase Generator

Figure 2–63 shows two VFCs connected as a staircase generator. Figure 2–64
shows the waveforms involved. The one-shot and comparator of IC-1 provides the
timing function (similar to that of a standard 555 timer IC), and the switched-current
source of IC-2 charges a capacitor.

When power is first applied, both C_1 and C_2 are discharged, and C_1 begins to
charge through R_1. The voltage on C_1 increases until the IC-1 threshold is reached.
The timing is set by R_A and R_B. When the threshold is exceeded, the one-shot fires,
bringing pin 3 low and delivering a discrete charge to C_2. The voltage on C_2 rises by
an amount proportional to the charge, and the logic output discharges C_1 through D_3.

After the one-shot period of IC-1, the logic output (pin 3) goes high, allowing
C_1 to charge again. The cycle repeats until enough charges have been delivered to C_2
to bring the C_2 voltage above the IC-1 comparator threshold (set by R_X and R_Y).
When the threshold has been exceeded, the IC-1 one-shot fires, bringing the logic
output low, and discharging both C_1 and C_2 through D_1 and D_2. After the one-shot period of IC-1, the logic output goes high, C_1 again begins charging through R_1, and a
greater cycle repeats itself.

The basic timing function is provided by the charge and discharge of C_1. This
timing sets the interval between chargings of C_2. Because C_2 does not have a dis-

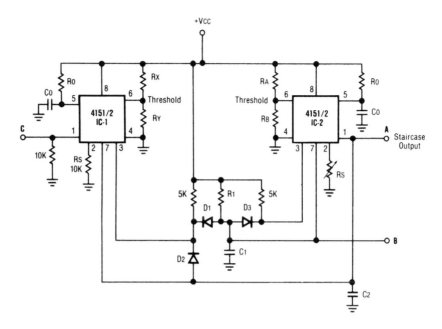

Figure 2–63. Staircase generator with two VFCs (*Raytheon Semiconductor Data Book,* 1994, p. 7-28)

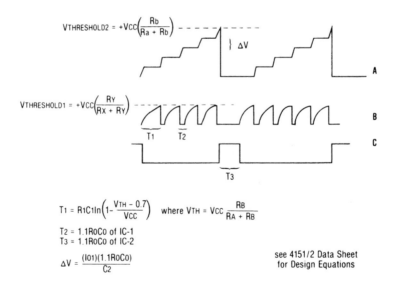

$$T_1 = R_1C_1\ln\left(1 - \frac{V_{TH} - 0.7}{V_{CC}}\right) \quad \text{where } V_{TH} = V_{CC}\,\frac{R_B}{R_A + R_B}$$

$T_2 = 1.1R_0C_0$ of IC-1
$T_3 = 1.1R_0C_0$ of IC-2

$$\Delta V = \frac{(I_{01})(1.1R_0C_0)}{C_2}$$

see 4151/2 Data Sheet
for Design Equations

Figure 2–64. Staircase generator waveforms (*Raytheon Semiconductor Data Book,* 1994, p. 7-29)

charge path, the voltage on C_2 increases in steps until the threshold is reached. A low-input current buffer on that output might be necessary, depending on circuit application.

2.17 Signal Conditioning

Both VFCs and FVCs might require signal-conditioning circuits at the input. This section describes three such circuits.

2.17.1 Transducer-Bridge Inputs

Figure 2–65 shows how the outputs of low-level bridge-type transducers can be applied to a VFC input using a low-drift instrumentation amplifier. The required excitation voltage for the bridge is provided by the 7.3-V reference output from the 4153. (Amplifier A1 and the transistor regulate the 7.3-V reference.) Changes in the value of transducer X cause changes in the differential output voltage (V_{DIFF}) from the bridge, as expressed by:

$$V_{DIFF} = V_{REF}/2 - R/X + R (V_{REF}).$$

The differential output is amplified by A_2 and applied to the VFC. As a result, the VFC output frequency is proportional to the value of the transducer X resistance. This circuit can be altered (unipolar or bipolar, linear or nonlinear) as necessary.

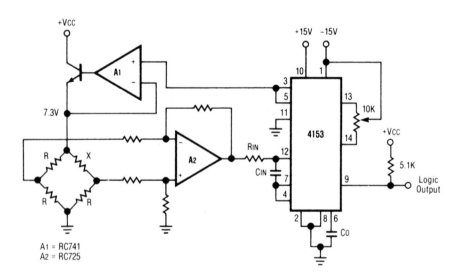

Figure 2–65. Signal conditioning for bridge-type transducers (*Raytheon Semiconductor Data Book,* 1994, p. 7-30)

2.17.2 Absolute-Value Circuit

Figure 2–66 shows an absolute-value circuit when bipolar inputs must be applied to a VFC. When V_{IN} is positive, the op amp is effectively out of the circuit. When the op-amp output goes low, D_2 turns off, and the signal passes (without attenuation) to the VFC input. When V_{IN} is negative, the op amp inverts the signal, and the VFC receives a positive input. (Diode D_2 is biased through zener D_1. This ensures that the polarity-indicating transistor is turned on.) The polarity output can be used to drive the sign bit of an interface or display.

2.17.3 FVC Signal Conditioning

Figure 2–67 shows the signal-conditioning circuits for an FVC. Figure 2–68 shows the waveforms. As discussed, FVCs generally require an input waveform with a sharp edge, and that the width of the pulse signal reaching the trigger input must be less than the period of the FVC one-shot. This prevents the one-shot from being re-triggered, which would cause an erratic, nonlinear output.

As an example, if the input is a sine wave, a Schmitt trigger or comparator should be used to square up the waveform before AC-coupling to the FVC. In a typical application, the time constant $C_A(R_A/R_B)$ should be kept less than 15 μs.

For applications with a square-wave input, the one-shot time should be kept less than the minimum period of the square wave. This prevents the input from interfering with the timing waveform and affecting linearity. This problem can be avoided by keeping C_o small.

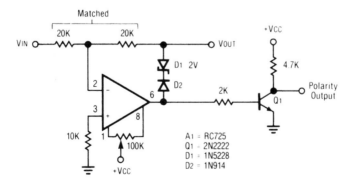

Figure 2–66. Absolute-value circuit for bipolar inputs (*Raytheon Semiconductor Data Book,* 1994, p. 7-31)

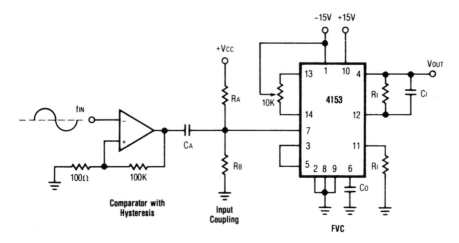

Figure 2–67. FVC input signal conditioning (*Raytheon Semiconductor Data Book,* 1994, p. 7-31)

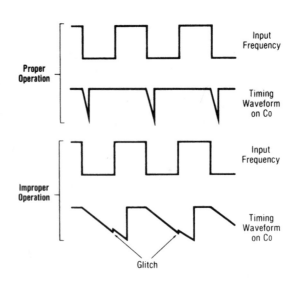

Figure 2–68.
FVC input signal-conditioning waveforms
(*Raytheon Semiconductor Data Book,* 1994, p. 7-31)

CHAPTER 3

Simplified Design with Analog Devices VFCs

This chapter is devoted to simplified-design approaches for Analog Devices VFCs (the AD537, AD650, AD651, and AD654). All of the general design information in Chapter 1 applies to the examples in this chapter. However, each voltage-frequency IC has special design requirements. The circuits in this chapter can be used immediately the way they are, or by altering component values, as a basis for simplified design of similar voltage-frequency conversion applications.

3.1 Description of AD537

Figure 3–1 shows the block diagram of the AD537. Figure 3–2 shows the basic circuit connections. The IC is designed for either single-supply or dual-supply operation at low voltages and low current (5 V and 1 mA), with the capability of driving high-voltage, high-current loads (36 V and 20 mA). The IC includes an accurate band-gap reference generator, a low-drift input amplifier capable of operating directly from millivolt signals, a precision current-controlled oscillator, and a high-current output stage. The circuit is complete, using low temperature-coefficient silicon-chromium thin-film resistors throughout, except for one external resistor and one capacitor.

The user is provided with a means for programming the full-scale (FS) input voltage from 100 mV to 10 V (or greater, depending on the positive supply voltage in some cases), and the FS frequency to any value less than 150 kHz. Either positive or negative input voltages can be accepted. The scale relationship is f = V/10RC. This ratio simplifies the choice of external components.

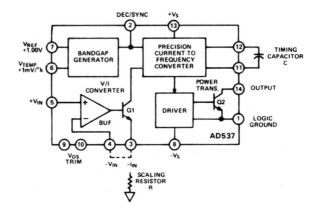

Figure 3-1.
Block diagram of AD537
(Analog Devices, *Applications Reference Manual,* 1993, p. 23-14)

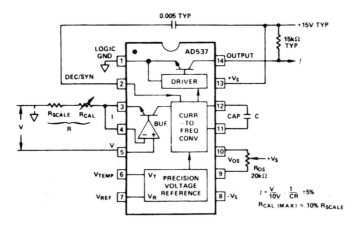

a. *14 Pin Dip*

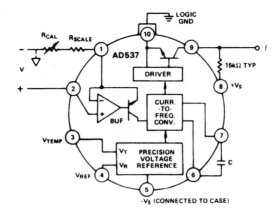

b. *10 Pin Metal Can*

Figure 3-2. Basic VFC circuit connections (Analog Devices, *Applications Reference Manual,* 1993, p. 23-15)

Linearity error is ±0.07% for 10-kHz full-scale over an 80-dB dynamic range. The output temperature is stable with temperature (typically ±50 ppm/°C excluding the effects of external components) and supply (typically ±0.01%/V from 5 to 36 V). The input amplifier (which actually functions as a voltage-to-current converter) has a typical drift of only ±1μV/°C when nulled. This permits operation directly from such low-level transducers as strain-gauges, thermocouples, current-shunts, and so forth, but with 250-M input resistance to positive signals. The output stage (an open-collector NPN) can sink up to 20 mA with a saturation voltage less than 0.4 V, and can withstand a supply of 36 V.

The logic-common terminal can be connected to any level between (or –VS) and 4-V below +VS, permitting easy interface with any digital logic family. The high-current capability permits LEDs, long cables, or up to 12 TTL loads to be driven directly. The square-wave output has several advantages. Internally, the power dissipation is essentially independent of frequency, so that self-heating effects do not cause linearity errors. Externally, the average level of the output is constant. This is useful in AC-coupled links, and allows operation as a phase-locked loop in FVC and other applications.

The IC has two auxiliary outputs. First, there is a fixed voltage of 1.00 V, generated by the primary band-gap reference circuit. This is useful in many applications. For example, the reference voltage can be used to power a resistive transducer, or to provide the reference for a digital-to-analog converter. The second auxiliary output is scaled +1.00 mV/°K. This output can be used as a reliable temperature-frequency converter. The self-heating error can be kept to less than 1°K using a 5-V supply and light output loading. When used with the fixed reference output, offset scales such as Celsius or Fahrenheit can be accommodated.

3.1.1 Basic Operating Theory

The key to accurate operation of the VFC is the current-to-frequency converter shown in Figs. 3–1 and 3–2. The circuit is essentially a multivibrator requiring only one timing capacitor C that is charged in push-pull fashion. This results in a large voltage across the capacitor, even at low supply voltages, and provides improved timing accuracy (low cycle-to-cycle jitter). The square-wave output is generally more useful than the typical narrow pulse generated by some charge-dispensing converters. The multivibrator can be operated with control currents from 100 nA to 2 mA (with proper biasing).

The basic circuit has a temperature-coefficient of 300 ppm/°C (at all values of control current). The band-gap reference supplies an exactly proportioned temperature-compensation voltage to the multivibrator, in addition to bias for internal operation of the chip, a fixed reference output, and the temperature-proportional output.

The output driver is also driven by the square-wave output from the multivibrator. This configuration provides a floating drive-current to output transistor Q2. (In reality, Q2 is formed by two transistors connected in a thermally symmetric layout for minimal interaction with the remaining circuits.) Transistor Q2 has a very low saturation voltage, combined with low on-voltages and a well-defined current limit.

The sync input permits the multivibrator to be slaved to a master clock, if desired, and controls the state of the square-wave output.

The input stage is an op-amp buffer used to convert the applied input voltage to a control current for Q1. This current is optimally 1 mA at the input corresponding to the maximum output frequency. By strapping pins 3 and 4 (as shown in Fig. 3–1), the input voltage can be impressed across the external scaling resistor R, which is chosen to provided the needed transconductance for the application. For example, for a full-scale input voltage of 2.5 V, the optimum resistor value is typically 2.5 k.

The +VIN terminal of the op-amp offers a high input resistance (250 M) with a bias current of about 100 nA. Drifts of about $1\mu V/°C$ can be obtained, even with input resistances of 1 k. As a result, the IC can accommodate millivolt signals without a pre-amplifier. The input configuration is also quasi-differential so that ground-loop errors can be avoided with the proper choice of signal connections.

The input common-mode range extends from 4-V below +VS (that is, from +11 V for a +VS of 15 V) down to –VS. As a result, inputs down to ground potential can be accepted, even when operating from a single-supply voltage. Negative inputs (voltages or currents) can be accepted by fixing the voltage on the +VIN terminal (usually to ground potential) and driving the –VIN and Q1-emitter terminals.

The internal voltage of about 1.22 V from the reference is divided by resistors to provide an output of 1 V to external circuits. The reference output terminal (V_{REF}) from the internal reference has a finite resistance of about 380 ohms. This can cause possible loading when the voltage is used to drive external circuits. Also, the temperature-proportional output (V_{TEMP}) of the reference has a source resistance of about 900 ohms, so loading must be considered. When the IC is used in the thermometer mode, the loading on the op-amp input is negligible.

Pins 9 and 10 of the 14-pin version (Fig. 3–2a) provide for trimming of the offset voltage of the input op amp to exactly zero, using an external pot with the slider connected to +VS. However, in many cases, the low initial offset (2mV guaranteed for the AD537K) eliminates the need for trimming (it amounts to a zero-error of only 0.02% for a 10-V input) and the IC operates properly without the pot. (Note that no damage will result to the circuit if pin 9 or pin 10 accidentally connects to –VS. Offset trimming is not available on the 10-pin can version of the IC, as shown in Fig. 3–2b.)

3.2 Applications for AD537 and AD654

This section describes a number of voltage-frequency converter applications using the AD537 and AD654.

3.2.1 Basic VFC Operation

Figure 3–2 shows the basic connections for an AD537 VFC. Note that the logic ground is strapped to the supply ground. The value of R is chosen so that the full-scale input voltage produces a current of 1 mA, as discussed. For example, for a 0 to +10-V input, R is 10 k. The value of C is set by the equation shown in Fig. 3–2a. For a 10-V FS input, the equation is simplified to: $C = 1/(CR)$. Typically, $C = 0.01$ µF for 10-kHz FS (1 Hz/mV), or 1000 pF for 100-kHz FS (10 Hz/mV), assuming an R of 10 k.

To calibrate, begin with 0 V input and adjust the optional offset pot (pins 9 and 10) for zero out. (A scope at the output will show when the IC stops oscillating and produces zero out.) Then apply the FS voltage input and adjust R_{CAL} for an output of 100 kHz. When FS is small, adjustment of R_{CAL} can affect offset, so it might be necessary to readjust the offset pot at pins 9 and 10. Typical input-voltage drift (after offset nulling) is 1 µV/°C.

3.2.2 VFC for Negative Inputs

Figure 3–3 shows the basic operating circuit for negative inputs. Note the similarity to Fig. 3–2.

3.2.3 VFC with Scale Adjustment for Current Inputs

Figure 3–4 shows a circuit that is similar to the circuit of Fig. 3–2, except that the input signal is in the form of a negative current. Capacitor C is selected to be 5% below the normal value. With R2 in the mid-position, the output frequency is given by:

$$F = I/(10.5 \times C)$$

where F is in kHz, I is in mA, and C is in µF.

For example, for an FS frequency of 10 kHz at an FS input of 1 mA, C = 9500 pF. To calibrate, apply the FS input current and adjust R2 for the correct FS output.

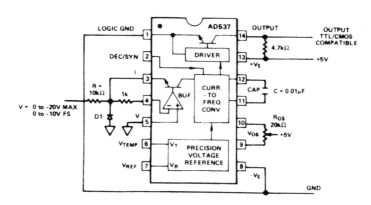

Figure 3–3. VFC for negative inputs (Analog Devices, *Applications Reference Manual*, 1993, p. 23-16)

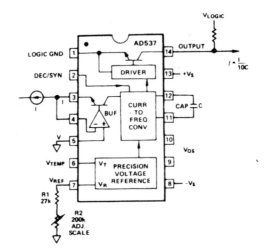

Figure 3–4.
VFC with scale adjust-
ment for current inputs
(Analog Devices, *Applications Reference Manual,*
1993, p. 23-16)

3.2.4 VFC with Digital Interfacing

Figure 3–5 shows the connections for interfacing the AD537 with various logic families and components. The required logic-common voltage, logic-supply voltage, pull-up resistor, and –VS supply are shown in the table. In the TTL mode, up to 12 standard gates (20 mA) can be driven at a maximum low voltage of 0.4 V.

3.2.5 VFC with Two-Wire Data Transmission

Figure 3–6 shows the connections for operating the AD537 at the remote end of a single wire-pair. The PNP transistor converts current modulation into a voltage sig-

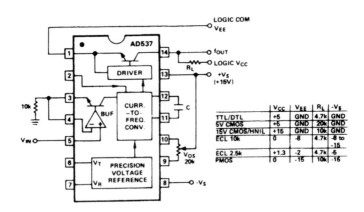

Figure 3–5. VFC with digital interfacing (Analog Devices, *Applications Reference Manual,* 1993, p. 23-16)

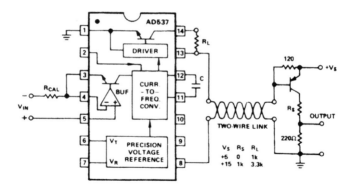

Figure 3–6. VFC with two-wire data transmission (Analog Devices, *Applications Reference Manual,* 1993, p. 23-17)

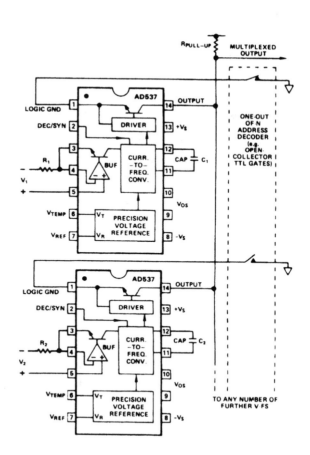

Figure 3–7.

VFC with signal multiplexing (Analog Devices, *Applications Reference Manual,* 1993, p. 23-17)

nal suitable for driving digital logic. The wire-pair line supplies power to the VFC. Using the values shown, the supply current through the lines is:

	Output off	*Output on*
Zero scale	1.2 mA	5.2 mA
Full scale (1 mA)	3.5 mA	7.5 mA

Approximately 500 mV of variation appears on the remote end of the supply line, but does not affect operation of the VFC.

3.2.6 VFC with Signal Multiplexing

Figure 3–7 shows the connections for multiplexing the outputs of several VFCs. In this circuit, all VFCs are operating continuously, but only the device having the LOGIC COMMON pin grounded (through the open-collector decoder or other digital-switching element) transmits an output.

3.2.7 VFC as an ADC

Figure 3–8 shows a VFC used as an analog-to-digital converter (ADC). Refer to Section 3.4 for further information on VFCs used as ADCs. Using the values

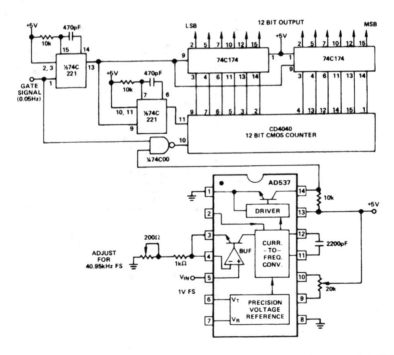

Figure 3-8. VFC as an ADC (Analog Devices, *Applications Reference Manual,* 1993, p. 23-19)

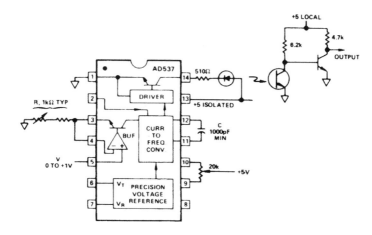

Figure 3–9. VFC with opto-coupling (Analog Devices, *Applications Reference Manual,* 1993, p. 23-20)

shown in Fig. 3–8, the circuit generates a binary output of 111111111111 (decimal 4095) and the first bit occurs for an input of 244 μV.

To calibrate, start by adjusting the offset pot as described for the circuit of Fig. 3–2. Then adjust the input for an output frequency (at pin 14) of 40.95 kHz as shown.

3.2.8 VFC with Opto-Coupling

Figure 3–9 shows a VFC connected for opto-coupling to a transmission line. The output pulses at pin 14 are amplified by the circuit to a level suitable for use with a fiberoptic transmitter or directly to a metal line.

3.2.9 Photodiode Preamplifier for VFC Output

Figure 3–10 shows a photodiode preamplifier suitable for converting pulses from a VFC and transmitted over fiberoptic cable into a logic output. The circuit has sufficient bandwidth to accept optical inputs up to 20 kHz.

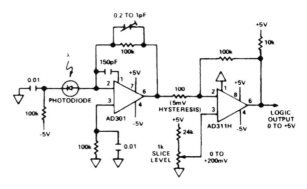

Figure 3–10.
Photodiode preamplifier for VFC output (Analog Devices, *Applications Reference Manual,* 1993, p. 23-20)

3.2.10 VFC with Bipolar Input

Figure 3–11 shows a VFC connected to accept bipolar inputs. To calibrate, set the input to zero and adjust R1 for an output frequency of 10 kHz. Then apply an input and adjust R2 for an output of 18 kHz.

3.2.11 VFC with Bipolar Input and a Stable Reference

Figure 3–12 shows a circuit similar to that of Fig. 3–11, but with an AD589 voltage-reference added.

3.2.12 VFC with Bipolar Input (Absolute Value)

Figure 3–13 shows a VFC that accepts bipolar inputs, but has double the range of linear operation (compared to the circuits of Figs. 3–11 and 3–12). Selection of values for R and C is the same as for the Fig. 3–2 circuit, as is calibration. However, the Fig. 3–13 circuit also requires that the op-amp offset be nulled by R8. The Fig. 3–13 circuit is scaled for an input of ±10 V full-scale. You can get best accuracy by nulling the input to the AD537 (with a digital voltmeter at pin 5). Note that if R2 and R3 are not properly matched, there might be a reversal at the input.

3.2.13 VFC with Thermocouple Interface

Figure 3–14 shows a circuit used to indicate temperature in °C on a counter/display with a 100-ms gate width. The VFC must deliver 7 kHz for an input of 53.14 mV. (The output of a Chromel-Constantan Type C thermocouple, with a reference junction of 0°C, varies from 0 to 53.14 mV over the temperature range of 0 to +700°C with a slope of 80.678 µV/degree over most of the range, and with some nonlinearity from 0 to +200°C.)

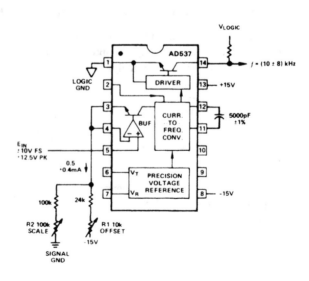

Figure 3–11.
VFC with bipolar input
(Analog Devices, *Applications Reference Manual*, 1993, p. 23-20)

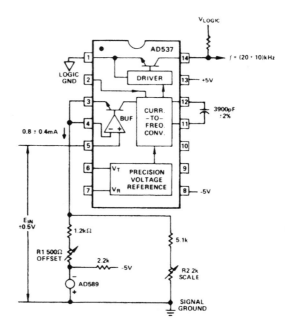

Figure 3–12.
VFC with bipolar input
and a stable reference
(Analog Devices, *Applications Reference Manual,*
1993, p. 23-20)

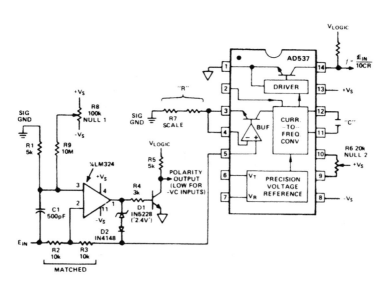

Figure 3–13. VFC with bipolar input (absolute value) (Analog Devices, *Applications Reference Manual,* 1993, p. 23-21)

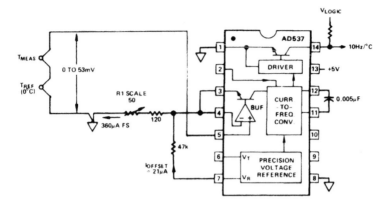

Figure 3–14. VFC with thermocouple interface (Analog Devices, *Applications Reference Manual,* 1993, p. 23-21)

The circuit provides the greatest accuracy from +300 to +700°C. To calibrate, raise the thermocouple to a known reference temperature (preferably near +500°C, halfway between 300 and 700) and adjust R1 for the correct readout on the counter/display. The error should be within ±0.2% over the range 400 to 700°C.

3.2.14 VFC with Strain-Gauge Input

Figure 3–15 shows a VFC connected to accept a strain-gauge input. The circuit is calibrated to generate a scale of Hz-per-microstrain (100 kHz at the assumed FS value). To calibrate, adjust R1 for zero offset, then apply an FS input (100 mV) and adjust R2 for 100 kHz at the output.

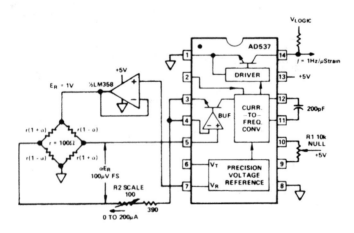

Figure 3–15. VFC with strain-gauge input (Analog Devices, *Applications Reference Manual,* 1993, p. 23-22)

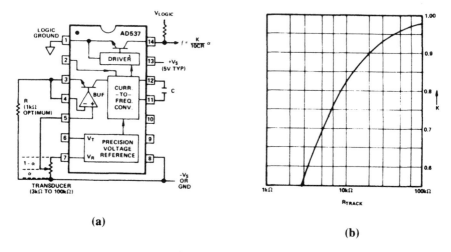

(a)

(b)

Figure 3–16. VFC with resistive-transducer interface (Analog Devices, *Applications Reference Manual,* 1993, p. 23-22)

3.2.15 VFC with Resistive-Transducer Interface

Figure 3–16 shows a VFC connected to accept resistive-transducer inputs. (Such transducers include linear displacement, rotary servo pot, level, light-comparator with photo-resistors, and so forth.) As shown by the equation, the output frequency depends on tranducer resistance R, capacitance C, and a constant K (Fig. 3–16b).

3.2.16 VFC with Resistive-Transducer Interface (Linear-Period)

Figure 3–17 shows a VFC similar to that of Fig. 3–16, except with linear-period control. That is, linear motion of the transducer-pot slider results in linear control of the period, rather than the frequency, of the output (frequency is inversely pro-

Figure 3–17.
VFC with resistive-transducer interface (linear-period) (Analog Devices, *Applications Reference Manual,* 1993, p. 23-23)

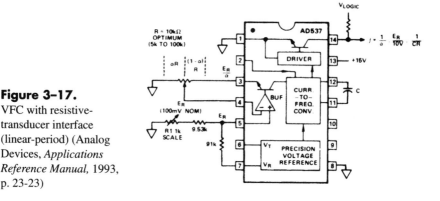

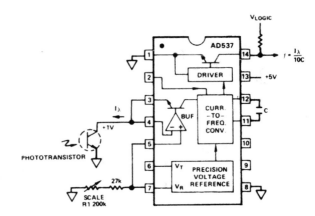

Figure 3–18.
VFC with phototransistor interface (Analog Devices, *Applications Reference Manual*, 1993, p. 23-22)

portional to time period). Unlike the circuit of Fig. 3–16, the circuit of Fig. 3–17 can be calibrated for a given output frequency.

3.2.17 VFC with Phototransistor Interface

Figure 3–18 shows a VFC with a phototransistor interface. As shown by the equation, the output frequency depends on capacitor C and the current generated by the phototransistor. The scale is calibrated to the desired output frequency (for a given phototransistor current) by R1.

3.2.18 VFC Connected for 4- to 20-mA Loop Operation

Figure 3–19 shows a VFC connected to convert instrumentation signals in the 4- to 20-mA format to a frequency format. (The output frequency is zero when the in-

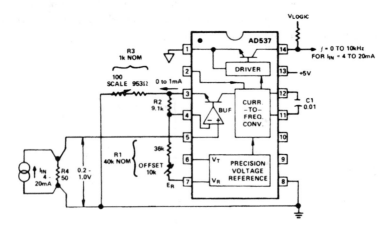

Figure 3–19. VFC connected for 4- to 20-mA loop operation (Analog Devices, *Applications Reference Manual,* 1993, p. 23-23)

put current is 4 mA.) Resistor R1 sets the offset for zero output (input 4 mA), and R3 sets the FS frequency of 10 kHz (input 20 mA).

3.2.19 VFC Connected for Self-Powered 4- to 20-mA Loop Operation

Figure 3–20 is similar to that of Fig. 3–19, except that the IC is powered by the 4- to 20-mA current, and the output is optically coupled to provide isolation. Using the values shown, R2 is adjusted for a 5-kHz output when the input current is 20 mA.

3.2.20 VFC as a Bell System Data Encoder

Figure 3–21 shows a VFC connected as an FSK (frequency-shift-keying) encoder suitable for Bell System modem communication (mark frequency of 1200 Hz and a space frequency of 2200 Hz). Resistor R3 sets the frequency of both mark and space outputs. The square-wave output must be filtered before transmission over a public telephone line.

3.2.21 VFC as an FVC

Figure 3–22 shows a VFC connected as an FVC. The circuit can lock onto any frequency from X to full-scale (10 kHz in this example) within four or five cycles. The DC output (taken from the filter outside the loop) is +1 V for a full-scale input.

To calibrate, set the V_{OS} pot to midscale, apply a frequency input of 10 kHz, and adjust R1 for +1 V. Then apply a 10-kHz input signal and trim V_{OS} for a 1-mV output. Retrim R1, as necessary, at 10 kHz.

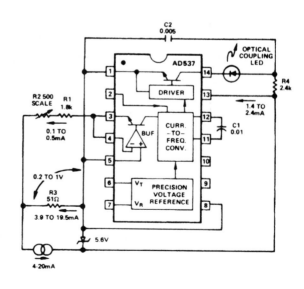

Figure 3–20.

VFC connected for self-powered 4- to 20-mA loop operation (Analog Devices, *Applications Reference Manual*, 1993, p. 23-24)

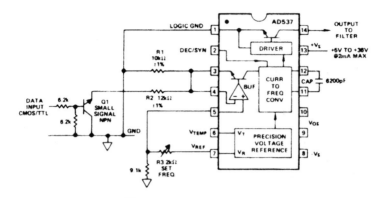

Figure 3–21. VFC as a Bell System data encoder (Analog Devices, *Applications Reference Manual,* 1993, p. 23-24)

3.2.22 VFC as a Bell System Data Decoder

Figure 3–23 shows a VFC connected as an FSK decoder for the Bell System modem communication (mark frequency of 1200 Hz and space frequency of 2200 Hz). No calibration is required. The operating range of this circuit is 800 to 2600 Hz.

3.2.23 VFC as a PLL (Phase-Locked Loop)

Figure 3–24 shows a VFC connected as a PLL. The waveforms are shown in Fig. 3–24b. The output at pin 14 is a noise-free square wave having exactly the same

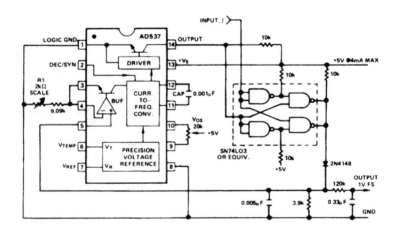

Figure 3–22. VFC as an FVC (Analog Devices, *Applications Reference Manual,* 1993, p. 23-25)

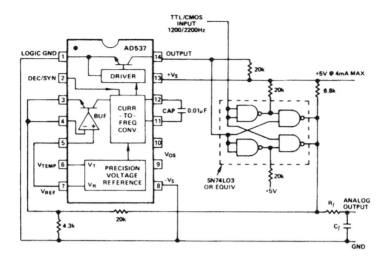

Figure 3–23. VFC as a Bell System data decoder (Analog Devices, *Applications Reference Manual,* 1993, p. 23-25)

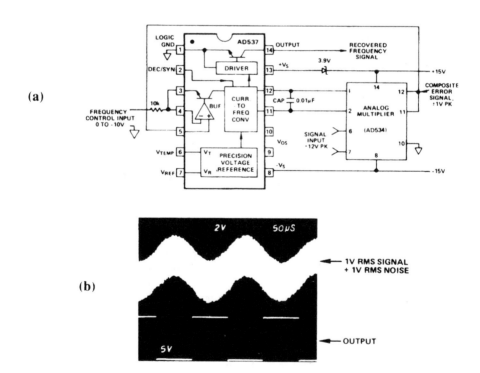

Figure 3–24. VFC as a PLL (phase-locked loop) (Analog Devices, *Applications Reference Manual,* 1993, p. 23-25)

frequency as the input signal at pins 6/7 of the analog multiplier. Frequency control is applied at pin 4 and no adjustments are required.

3.2.24 VFC as an Analog Integrator

Figure 3–25 shows a VFC connected as an analog integrator. Figure 3–25b shows the waveforms (output of integrator). Figure 3–25c shows the integration equation. Figure 3–25d shows the values of C and R for various times and frequencies, using a +1-V analog input. Note that the first two bits of the 4080 counter provide a prescaler.

3.2.25 VFC as an Analog Divider

Figure 3–26 shows a VFC connected as an analog divider (with frequency output). That is, the output frequency is proportional to the ratio of the two input voltages V_D and V_N, representing the denominator and numerator. The output frequency depends on the values of R_1, R_3, and C. For an R_1 of 2.4 k, the output frequency is VN/VCR3. To adjust the denominator offset, connect the V_N and V_D inputs together and trim to maintain frequency independent of input voltage. Linearity of division is typically ±0.1%.

3.2.26 VFC as a Sound-Velocity Monitor

Figure 3–27 shows a VFC connected as a sound-velocity monitor. The circuit uses the temperature and reference output of the IC because the velocity of sound is related to temperature. The relationship is $V_S = (331.5 + 0.6T_C) = (167.6 + 0.6T_K)$, where V_S is sound velocity in m/s (meters per second), T_C is Celsius temperature, and T_K is Kelvin temperature.

Using the values shown, the voltage on pin 5 is 452.8 mV at 300°K, which is scaled by R4 to an output frequency of 347.6 Hz (corresponding to the velocity of sound at 300°K). As shown by the equation, R4 is adjusted so that the output is 1 Hz when the VFC monitors a sound velocity of 1 m/s.

3.2.27 Basic VFC with Square-Wave Output

Figure 3–28 shows the basic connections for an AD654 VFC. The circuit is similar in function to that of Fig. 3–2, but with some differences (such as a square-wave output for the VFC of Fig. 3–28). As shown by the equations, the output frequency is related to input voltage V_{IN} (or the value of input current I) and the values of R1 and C1. In practical circuits, R1 is made adjustable to provide a specific scale of input voltage to output frequency.

3.2.28 VFC with Positive Input

Figure 3–29 shows the AD654 VFC connected to convert positive inputs to a corresponding output frequency. Figure 3–29b shows an offset-trim bias network (if

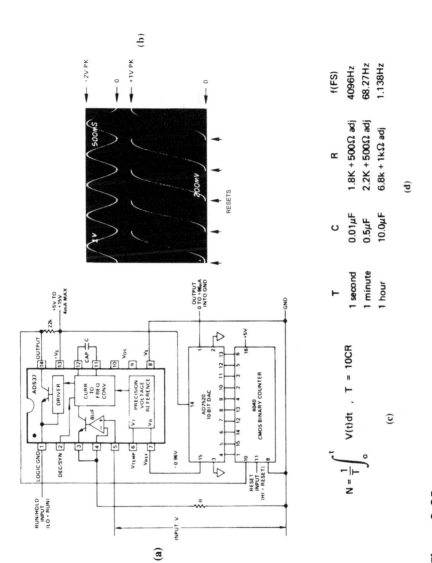

Figure 3-25.
VFC as an analog integrator (Analog Devices, *Applications Reference Manual*, 1993, p. 23-27)

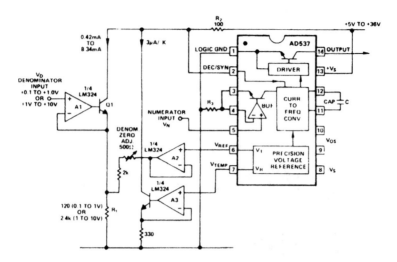

Figure 3–26. VFC as an analog divider (Analog Devices, *Applications Reference Manual,* 1993, p. 23-27)

required). Figures 3–29c and 3–29d show the values of R_1 and C_1 for various full-scale voltage input and output frequencies.

To calibrate, apply a known full-scale input voltage and adjust R_{CAL} for the desired FS output, using the values of Fig. 3–29e. If the offset circuit of Fig. 3–29b is used, begin calibration by applying a zero input (short the inputs together) and adjust the offset pot R3 for zero output (a scope that the output pin will show when the IC stops oscillating and produces zero out). It might be necessary to work between the FS and offset adjustments because they are interactive.

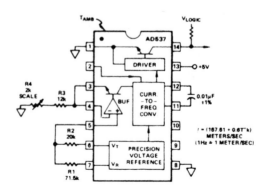

Figure 3–27.
VFC as a sound-velocity monitor (Analog Devices, *Applications Reference Manual,* 1993, p. 23-28)

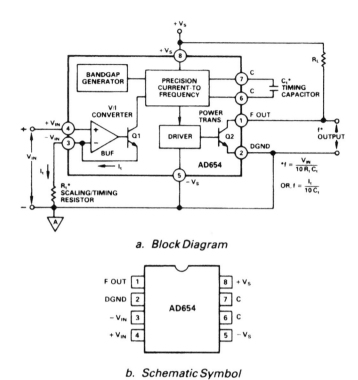

a. Block Diagram

b. Schematic Symbol

Figure 3–28. Basic VFC with square-wave output (Analog Devices, *Applications Reference Manual,* 1993, p. 23-31)

3.2.29 VFC for Negative-Input Operation (Square-Wave Output)

The circuit of Fig. 3–30 is similar to that of Fig. 3–29, except for the negative-input capability.

3.2.30 VFC with Negative Current Input

The circuit of Fig. 3–31 is similar to that of Fig. 3–29, except that the input signal is in the form of a negative current.

3.2.31 VFC with Standard Logic Interfacing

Figure 3–32 shows the connections for interfacing the AD654 with various logic families and components. The required logic common voltage, logic supply, pull-up resistor, and –VS supply are shown in the table. In the TTL mode, up to 12 standard gates (20 mA) can be driven at a maximum low voltage of 0.4 V.

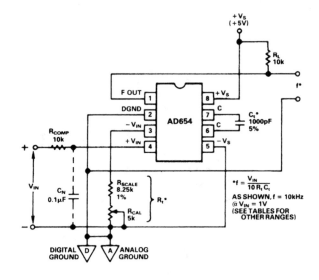

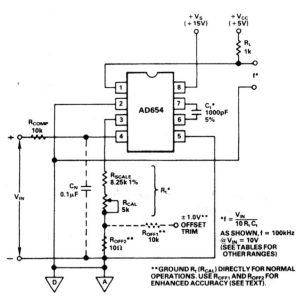

Figure 3-29.
VFC with positive input (Analog Devices, *Applications Reference Manual,* 1993, pp. 23-33–23-35)

(a)

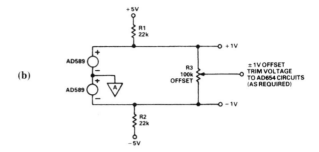

(b)

(c)

FS V$_{IN}$	R$_t$ FS I$_t$ = 100μA	R$_t$ FS I$_t$ = 1mA
100V*	1 meg	100k
10V	100k	10k
1V	10k	1k
100mV	1k	100Ω

NOTE
*Applies *only* to Figure 3.

(d)

FS f	C$_t$ FS I$_t$ = 100μA	C$_t$ FS I$_t$ = 1mA
≥1MHz**	*	≤100pF**
500kHz	*	200pF
250kHz	*	390pF
100kHz	*	1000pF
10kHz	1000pF	10000pF

Notes
*Not recommended, see text.
**"Exalted" operation, see text.

(e)

FS V$_{IN}$	FS Frequency 10kHz	100kHz	500kHz
10V	10V → 10kHz 1mV → 1Hz	10V → 100kHz 1mV → 10Hz	10V → 500kHz** 1mV → 50Hz*
1V	1V → 10kHz 1mV → 10Hz	1V → 100kHz 1mV → 100Hz	1V → 500kHz** 1mV → 500Hz*
100mV	100mV → 10kHz 1mV → 100Hz	100mV → 100kHz 1mV → 1kHz	100mV → 500kHz** 1mV → 5kHz*

Notes
*Adjust OFFSET (if used) as noted.
**Adjust FS cal as noted.

Figure 3-29.
Continued.

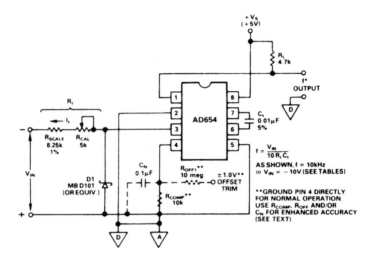

Figure 3–30. VFC for negative-input operation (square-wave output) (Analog Devices, *Applications Reference Manual,* 1993, p. 23-36)

With pin 2 grounded, pin 1 shorted to +15 V, and the oscillator running, the average power in the output stage is about 265 mW. The power is reduced to about one-third when the supply is +5 V. However, if +15 V is used and the output is in the on-state for long periods (low frequency), the peak dissipation is about 1260 mW. Also, the dissipation can cause heating (a special problem for the plastic package).

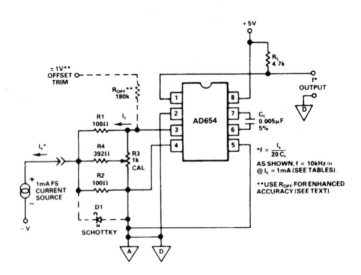

Figure 3–31. VFC with negative current input (Analog Devices, *Applications Reference Manual,* 1993, p. 23-37)

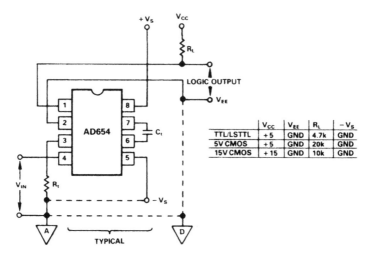

a. Standard Interfacing

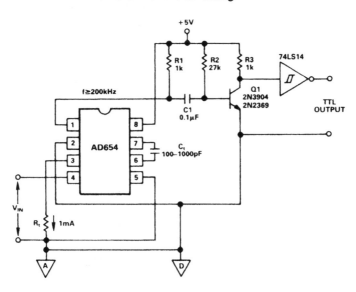

b. High-Speed TTL Output Buffer

Figure 3–32. VFC with standard logic interfacing (Analog Devices, *Applications Reference Manual,* 1993, p. 23-39)

3.2.32 VFC with Phantom Power

Figure 3–33 shows the connections to receive power from the system or component to which the VFC output is applied.

3.2.33 References for VFCs

Figure 3–34 shows several different reference configurations for VFCs. Such reference circuits can be used for bias, offset, or to establish scaling in a VFC application.

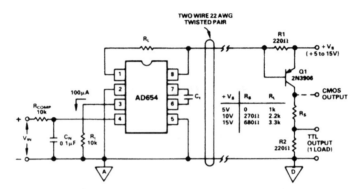

a. Simple Phantom Power Driver

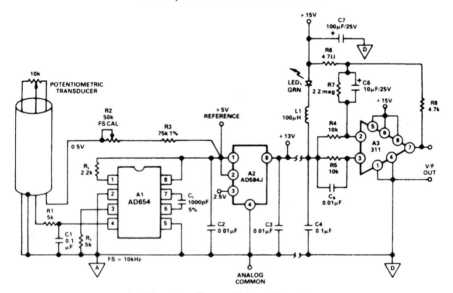

b. Phantom Power Driver/Regulator

Figure 3–33. VFC with phantom power (Analog Devices, *Applications Reference Manual,* 1993, p. 23-40)

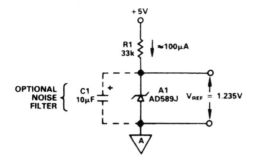

a. Basic Positive Bias Reference Source

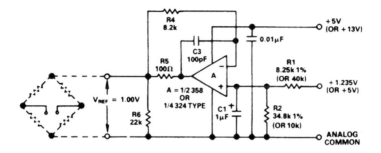

b. Buffered Low Power Reference

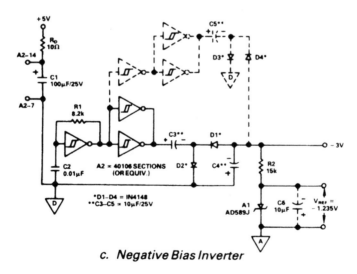

c. Negative Bias Inverter

Figure 3–34. References for VFCs (Analog Devices, *Applications Reference Manual,* 1993, p. 23-41)

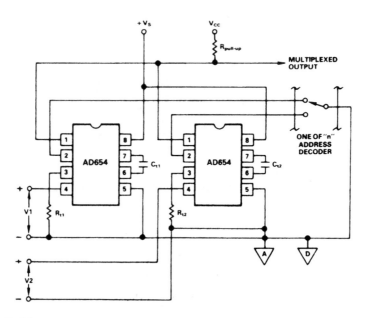

Figure 3–35. VFC with signal multiplexing (alternate) (Analog Devices, *Applications Reference Manual,* 1993, p. 23-41)

3.2.34 VFC with Signal Multiplexing (Alternate)

Figure 3–35 shows the connections for multiplexing the outputs of several AD654 VFCs. In this circuit, all VFCs are operating continuously, but only the device having the D_{GND} (pin 2, digital ground) pin grounded transmits an output.

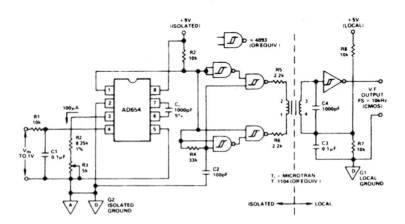

Figure 3–36. VFC with transformer signal isolation (Analog Devices, *Applications Reference Manual,* 1993, p. 23-42)

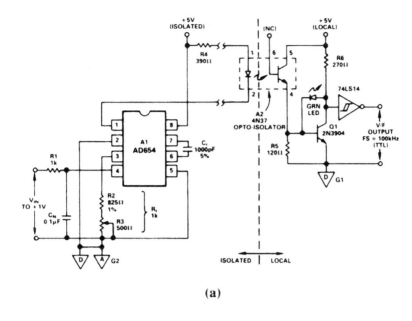

(a)

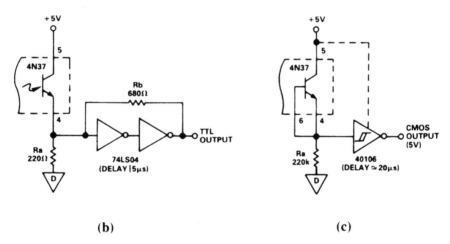

(b)   (c)

Figure 3–37. VFC with medium-speed opto-coupling (100 kHz) (Analog Devices, *Applications Reference Manual,* 1993, p. 23-42)

3.2.35 VFC with Transformer Signal Isolation

Figure 3–36 shows an AD654 VFC connected to provide an output signal to a system with a separate digital ground.

3.2.36 VFC with Medium-Speed Opto-Coupling (100 kHz)

Figure 3–37 shows the AD654 connected for opto-coupling to a transmission line. Figures 3–37b and 3–37c show alternate TTL and CMOS output buffers, respectively.

3.2.37 VFC with High-Speed Opto-Coupling (200–500 kHz)

Figure 3–38 shows the AD654 connected for opto-coupling to a transmission line, where the output frequency is substantially higher than 100 kHz. The circuit is useful up to about 1 MHz, but the greatest linearity is at about 200 kHz.

3.2.38 Single-Supply VFC with Bipolar Inputs

Figure 3–39 shows the AD654 connected for single-supply (+5 V) operation with bipolar (±10 V) inputs. To calibrate, apply a –10-V input and adjust R12 for an output of 10kHz. If the alternate R3 is used to compensate for rollover error (where there is a difference between equal-magnitude positive and negative inputs), trim R3 for a positive gain equal to negative gain. (A +10-V input should produce a 10-kHz output, as does a –10-V input.) The sign-bit output should be high, and the LED should be on, for a positive input.

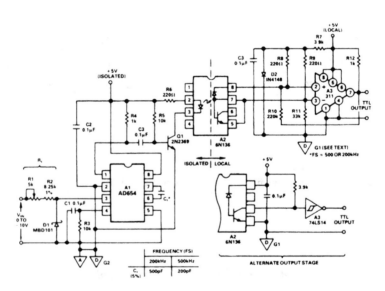

Figure 3–38. VFC with high-speed opto-coupling (200–500 kHz) (Analog Devices, *Applications Reference Manual,* 1993, p. 23-42)

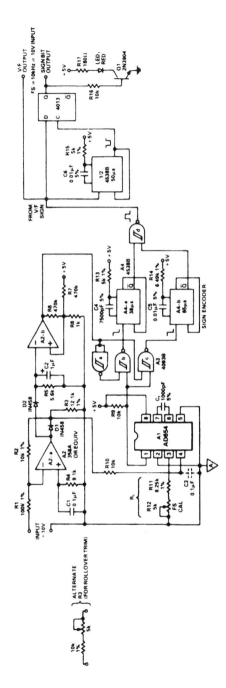

Figure 3–39.
Single-supply VFC with bipolar inputs (Analog Devices, *Applications Reference Manual*, 1993, p. 23-43)

3.2.39 Dual-Supply VFC with Bipolar Inputs

Figure 3–40 shows the AD654 connected for dual-supply (±5 V) operation with bipolar (±1 V or ±10 V) inputs. To calibrate, apply either ±10 V or ±1 V and adjust R4 for the correct output (using the frequency equation).

3.2.40 VFC with Linear Resistive-Transducer Interface

Figure 3–41 shows the AD654 connected to accept linear resistive-transducer inputs. With the values shown, adjust R2 so that the output frequency is 250 kHz (full-scale) when alpha is at unity.

3.2.41 VFC with Phototransistor Input

Figure 3–42 shows the AD654 connected to accept phototransistor inputs. If necessary, trim C_T (film trimmer) to produce the desired output frequency.

3.2.42 VFC with Photodiode Input

Figure 3–43 shows the AD654 connected to accept photodiode inputs. If necessary, trim C_T (film trimmer) to produce the desired output frequency. (Note: Figure 15 shown in the illustration refers to Fig. 3–42 in this book.)

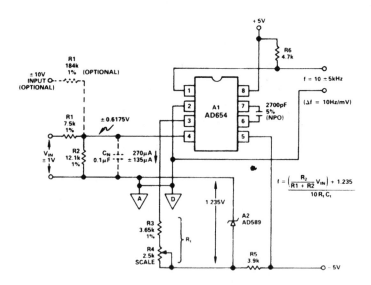

Figure 3–40. Dual-supply VFC with bipolar inputs (Analog Devices, *Applications Reference Manual*, 1993, p. 23-44)

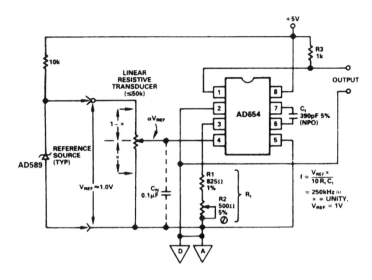

Figure 3–41. VFC with linear resistive-transducer interface (Analog Devices, *Applications Reference Manual,* 1993, p. 23-45)

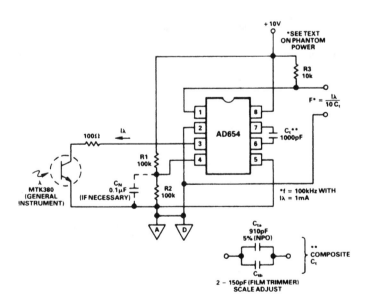

Figure 3–42. VFC with phototransistor input (Analog Devices, *Applications Reference Manual,* 1993, p. 23-45)

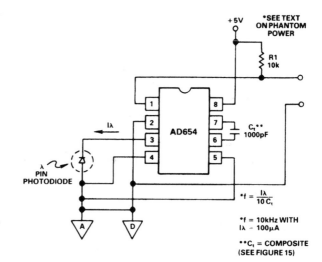

Figure 3–43.
VFC with photodiode
input (Analog Devices,
*Applications Reference
Manual,* 1993, p. 23-45)

3.2.43 VFC for 4- to 20-mA Loop Operation

Figure 3–44 shows the AD654 used to convert instrumentation signals in the 4-to 20-mA format to a frequency format. Resistor R2 sets the offset for zero output (when input is 4 mA), and R5 sets the full-scale frequency of 10 kHz (when the input is 20 mA). These adjustments are interactive.

3.2.44 Self-Powered VFC for 4- to 20-mA Loop

The circuit of Fig. 3–45 is similar to that of Fig. 3–44, except that the VFC is powered by the 4- to 20-mA current, and the output is optically coupled to provide

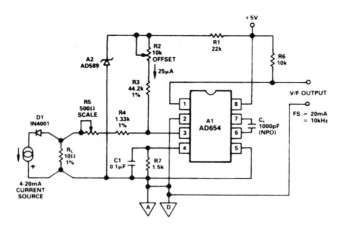

Figure 3–44.   VFC for 4- to 20-mA loop operation (Analog Devices, *Applications Reference Manual,* 1993, p. 23-47)

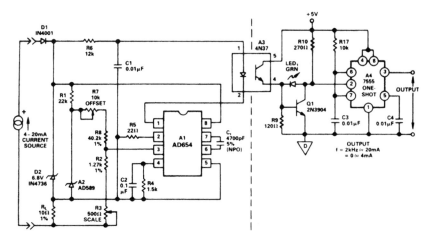

Figure 3–45. Self-powered VFC for 4- to 20-mA loop (Analog Devices, *Applications Reference Manual,* 1993, p. 23-47)

isolation. Resistor R7 sets the offset for zero output (input 4 mA), and R3 sets the full-scale frequency of 2 kHz (input 20 mA). These adjustments are interactive.

3.2.45 Digitally-Tuned Switched-Capacitor Filter

The circuit of Fig. 3–46 uses the AD654 as a single-supply, digitally program-mable clock source for a switched-capacitor filter. Using the values shown, the VFC output is a 10-V square wave (f_{CLK}), which satisfies the clocking requirements of the MF10C filter. In this circuit, the MF10C is programmed for bandpass operation. Re-sistor R11 is adjusted so that the maximum 12-bit input produces a clock of 100 kHz. Resistor R4 sets the level of the signal to be filtered.

3.2.46 VFC as a PLL (Alternate)

Figure 3–47 shows the AD654 connected as a PLL (phase-locked loop). Figure 3–24b shows the waveforms. The output is a noise-free square wave having exactly the same frequency as the input signal. Resistor R11 is adjusted for exactly 0.5 V at the VFC input (for full-scale). Resistance R_D and capacitance C_L set the dynamic range and/or speed. The waveforms of Fig. 3–47b are for a C_L of 0.1 µF and and R_D of 500 ohms. With the loop adjusted for a wide dynamic range ($C_L = 1$ µF, $R_D = 160$), the circuit will maintain a frequency and phase lock from 100 kHz down to about 500 Hz.

3.2.47 VFC as a Bus Monitor

Figure 3–48 shows the AD654 connected as a monitor for a –5-V bus. Resistor R4 is adjusted to provide a scale of 1Hz per mV. (The output frequency should be 5000 Hz when the bus is at 5 V.)

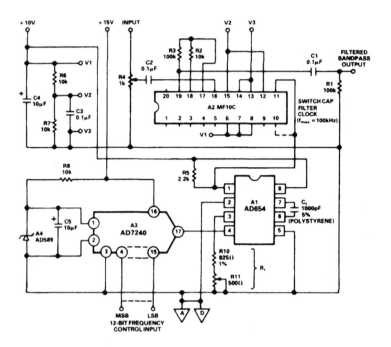

Figure 3–46. Digitally-tuned switched-capacitor filter (Analog Devices, *Applications Reference Manual,* 1993, p. 23-48)

3.2.48 VFC as a Negative Supply-Current Monitor

Figure 3–49 shows the AD654 connected as a monitor for a negative supply current (–20 V in this example). As shown by the table, the current limit to be monitored is set by the value of R_S. Resistor R3 sets the corresponding scale. For example, to monitor a maximum current of 1 A, use 0.1 for R_S and adjust R3 for a scale of 1 Hz per mA. (The output frequency should be 1000 Hz when the bus current is 1 A.)

3.2.49 VFC as a Sine-Wave-Averaging AC Monitor

Figure 3–50 shows the AD654 connected as a monitor for an alternating current (120-V line current in this case). As shown by the table, the current limit to be monitored is set by the value of R_S. Resistor R7 sets the corresponding scale. For example, to monitor a maximum current of 10 A, use 0.01 for R_S and adjust R7 for a scale of 0.1 Hz per mA. (The output frequency should be 1000 Hz when the line current is 10 A.)

3.2.50 Bipolar VFC with Ultra-High Input Impedance

Figure 3–51 shows the AD654 connected to accept bipolar inputs. The circuit input is set by the value of R_{IN}. For the calibration shown, apply –5 V and trim R2 for

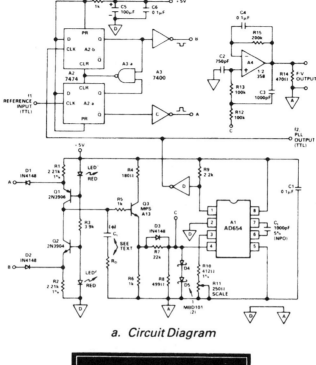

a. Circuit Diagram

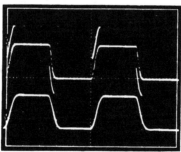

b. AD654 PLL Performance (Fast Response Mode)
Upper Trace: Phase Detector Output (Point "C")
Lower Trace: F/V Filtered Output

Scales: 0.2V/div, 500µs/div
Source: 50/100kHz FSK @ 400Hz Rate
Condition: $C_L = 0.1µF, R_d = 500Ω$

Figure 3–47. VFC as a PLL (alternate) (Analog Devices, *Applications Reference Manual,* 1993, p. 23-49)

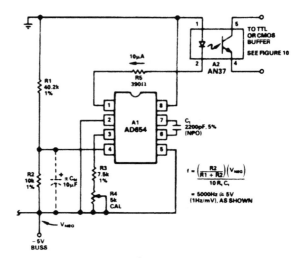

Figure 3-48.
VFC as a bus monitor
(Analog Devices, *Applications Reference Manual,*
1993, p. 23-53)

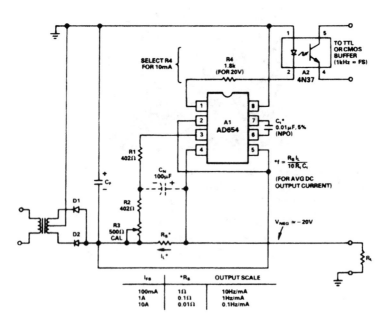

Figure 3-49. VFC as a negative supply-current monitor (Analog Devices, *Applications Reference Manual,* 1993, p. 23-53)

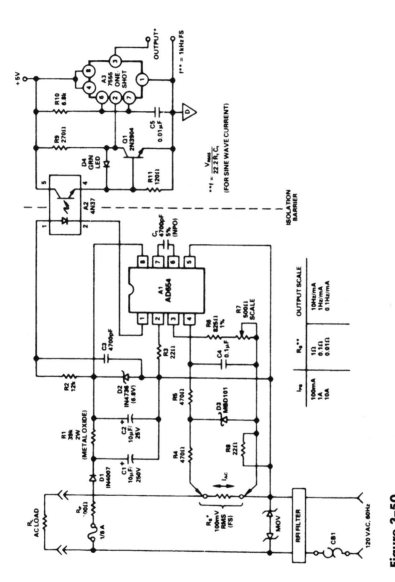

Figure 3-50.
VFC as a sine-wave-averaging AC monitor (Analog Devices, *Applications Reference Manual*, 1993, p. 23-53)

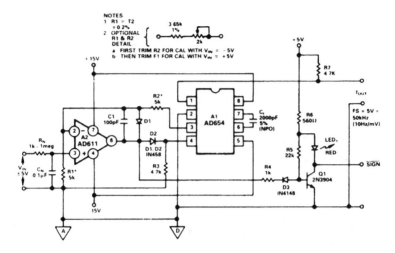

Figure 3–51. Bipolar VFC with ultra-high input impedance (Analog Devices, *Applications Reference Manual,* 1993, p. 23-54)

a 50-kHz output. Then apply +5 V and trim R1 for a 50-kHz output. The $\overline{\text{SIGN}}$ output should be high, and the LED should be on, for a positive input.

3.2.51 VFC with Output Doubling

Figure 3–52 shows the AD654 connected to convert the normal square-wave output into a pulse train (effectively doubling the output frequency) but still preserving the low-frequency linearity of the VFC. In this circuit, the ratio of R1 to R2 scales the 0 to +10-V input down to 0 to +1 mV at pin 4. R4 is adjusted to produce a 400-kHz full-scale output for a +10-V input.

3.2.52 VFC with Output Doubling (2 MHz)

The circuit of Fig. 3–53 is similar to that of Fig. 3–52, except with a much greater output-frequency capability. In the circuit of Fig 3–53, R2 is adjusted to produce a 2-MHz full-scale output for a +1-V input.

3.3 Using a VFC as an FVC

This section describes how a VFC can be used as an FVC. The AD650 is used as an example.

3.3.1 Basic FVC Operation

Figure 3–54 shows the major components of an FVC. Figure 3–55 is a simplified schematic of a VFC connected as an FVC. Figures 3–56 represents the current delivered to the integrator shown in the basic and simplified circuits.

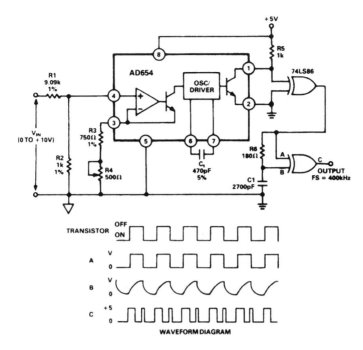

Figure 3–52. VFC with output doubling (Analog Devices, *Applications Reference Manual,* 1993, p. 23-56)

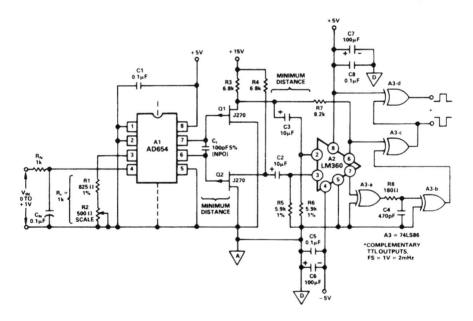

Figure 3–53. VFC with output doubling (2 MHz) (Analog Devices, *Applications Reference Manual,* 1993, p. 23-56)

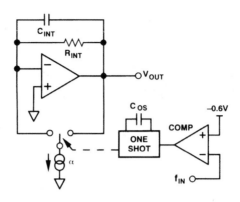

Figure 3–54.

Major components of an
FVC (Analog Devices,
*Applications Reference
Manual,* 1993, p. 23-59)

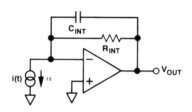

Figure 3–55.

Simplified schematic of a
VFC connected as an
FVC (Analog Devices,
*Applications Reference
Manual,* 1993, p. 23-59)

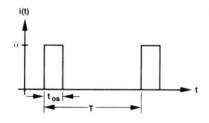

Figure 3–56.

Current delivered into an
integrator of FVC (Ana-
log Devices, *Applications
Reference Manual,* 1993,
p. 23-59)

The basic circuit consists of a comparator, a one-shot with switch, a constant-current source, and a lossy integrator. When the input signal crosses the threshold at the comparator input, the comparator triggers the one-shot. In turn, the one-shot controls a single-pole double-throw switch that directs the current source to either the summing junction, or the output, of the lossy integrator.

When the one-shot is in the ON state, current is injected into the integrator input, causing the integrator output to rise. When the one-shot period has passed, the current is steered to the integrator output. Because the output is a low-impedance node, the current has no effect on the circuit, and is effectively turned off. During this time, the output falls because C_{INT} discharges through R_{INT}. When there is a constant triggering applied to the comparator, C_{INT} charges to a relatively steady value and is maintained by constant charging and discharging. The charged stored on C_{INT} is unaffected by loading because of the op amp's low output impedance.

3.3.2 A Practical FVC Circuit

Figure 3–57 shows a practical version of the circuits in Figs. 3–54 and 3–55. The circuit of Fig. 3–57 is best suited to those applications where the input signal has slower edges than the one-shot time delay. In such applications, the comparator has a tendency to "double trip," possibly resulting in inaccurate results. (The AD650 data sheet has a simpler version of this circuit, but requires input signals with fast edges, such as a TTL signal.)

The current $i(t)$ shown in Fig. 3–55 can be thought of as a series of charge packets delivered at frequency $f_{IN} = (1/T)$, with constant amplitude (α) and duration (t_{os}) as shown in Fig. 3–56. The average value of the input current is found by dividing the area of the current $i(t)$ by the period T. The DC component of the output voltage is found by scaling the average input current by the feedback resistor R_{INT}. The average voltage output becomes a linear function of frequency when f_{IN} is substituted for $(1/T)$, or $V_{outavg} = t_{os} \times R_{INT} \times (\alpha) \times f_{IN}$.

The relationship between the average output voltage and input frequency is a function of the one-shot time constant and the feedback resistor, but not of the integration capacitor. This is because C_{INT} is an open circuit to DC. From a simplified-design standpoint, the most practical way to trim the full-scale voltage is to include a trim pot in series with R_{INT}, such as R3 in series with R1 (Fig. 3–57). Typically, a 30% trim range is required to absorb errors associated with t_{os} and (α).

3.3.3 Simplified Design Procedure

Start by assuming that the one-shot ON time is some fraction of the total input period, which is the time that the circuit integrates the current signal (α). The output ripple can be minimized by allowing the current source to be on during the majority

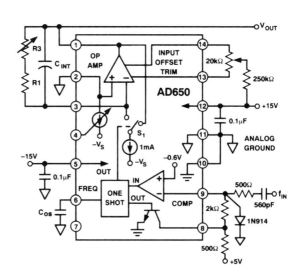

Figure 3–57.
Practical FVC circuit (Analog Devices, *Applications Reference Manual,* 1993, p. 23-63)

of the total input period. This is done by choosing the one-shot time constant so that it occupies almost the full period of the input signal when this period is at its minimum (or when the input frequency is maximum, f_{max}).

To allow for component tolerances at f_{max}, make t_{os} approximately equal to 90% of the minimum period. With t_{os} established, the value of the one-shot timing capacitor, C_{os}, can be determined from the following equation:

$$C_{os} = t_{os} - 3 \times 10^{-7} \text{ sec}/6.8 \times 10^3 \text{sec/F}$$

where t_{os} is in seconds and C_{os} is in farads.

For maximum linearity performance, use a low dielectric absorption capacitor for C_{os}. When C_{os} is known, the integration resistor R_{INT} (or the total of R1 and R3 in Fig. 3–57) is determined from the V_{outavg} equation, because t_{os}, (α), f_{IN}, and V_{out} are known.

Integration resistor C_{INT} is selected by determining the response time of the device being measured. For example, if the frequency signal to be measured is taken from a mechanical device such as an aircraft turbine shaft, the momentum of the shaft and the blades should be used to determine the response time. The time constant of the FVC is then set to match the time constant of the mechanical system. (The FVC time constant can be set somewhat lower, depending on the desired total response time of the mechanical and electrical system.) Allow several time constants (N) for the FVC to approach the final value.

Use the following expression to determine the value of C_{INT},

$$C_{INT} = \text{Mechanical Response Time}/N \times R_{INT}$$

where N is the number of time constants chosen to provide adequate settling time. The table in Fig. 3–58 can be used to determine the number of time constants required for a given settling accuracy, and the number of bits.

Note that a larger number of time constants gives a more responsive circuit, but will also increase the ripple at the FVC output. For simplified design, start with 8-bit

Figure 3–58.

Settling accuracy versus number of time constants (Analog Devices, *Applications Reference Manual,* 1993, p. 23-60)

# of Time Constants (N)	# of Bits	% Accuracy
4.16	6	1.6
4.85	7	0.8
5.55	8	0.4
6.23	9	0.2
6.93	10	0.1
7.62	11	0.05
8.30	12	0.024
9.00	13	0.012
9.70	14	0.006
10.4	15	0.003
11.0	16	0.0015

settling accuracy (0.4%), using N = 6 time constants, and increase or decrease N, depending on the ripple content.

The equation for determining ripple is long and complex. The graph of Fig. 3–59 shows the peak-to-peak ripple versus frequency for a typical application. The important point to remember is that ripple amplitude changes with frequency and is largest at the lowest frequency. For simplified design, the ripple amplitude must be below the voltage for 1/2 LSB, using an 8-bit conversion.

3.3.4 Design Example

Assume that the rpm of an automobile engine is to be monitored for use by an on-board computer. The rpm signal generated by the FVC is to be digitized with an 8-bit analog-to-digital converter (ADC). The engine range extends from 300 to 7,000 rpm. A 200-tooth flywheel converts rpm into pulses that range from 1,000 to 23,000 Hz (Fig. 3–59). The response time to a step change in throttle position is measured, and is found to be 400 ms in neutral. The goal is to design an FVC that will respond at approximately the same rate as the engine (or faster) and will have ripple that is undetectable by the ADC, which has a full-scale of 10 V for the 8 bits. The 1/2 LSB voltage for the ADC is 19.5 mV. Therefore, the ripple must be less than 19.5 mV. If you are not familiar with ADCs, read the author's *Simplified Design of Data Converters* (Butterworth–Heinemann, 1997).

Start by letting t_{os} be 0.9 times $(1/f_{max})$, or 0.9×43.5 μs = 39 μs.

Then find C_{os} using the equation of Section 3.3.3, resulting in a C_{os} of 0.0053 μF. This is not a practical value for polystyrene, but tantalum can be used with reduced linearity.

With C_{os} established, find the value of R_{INT} by rearranging the V_{outavg} equation of Sec. 3.3.2. Use an f_{max} of 23 kHz, a t_{os} of 39 μs, a current of 1 mA (the current-source value for the AD650), and a V_{outavg} of 10 V (the full-scale value of the 80-bit ADC), or

$$R_{INT} = 10 \text{ V}/1 \text{ mA} \times 39 \text{ μs} \times 23 \text{ kHz} = 11.14 \text{ k}$$

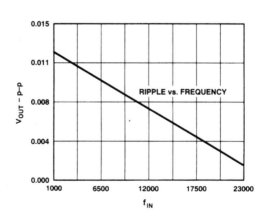

Figure 3–59.
Peak-to-peak ripple versus frequency (Analog Devices, *Applications Reference Manual*, 1993, p. 23-61)

Note that R_{INT} is the total of R1 and R3 in Fig. 3–57, and is the load seen by the AD650 op amp. If the calculated value is less than 1 k, try a different value of t_{os}.

Find the value of C_{INT} using the equation of Sec. 3.3.3. Use our 400 ms for the mechanical response time, 11.14 k for R_{INT}, and 6 for the number of time constants, or

$$C_{INT} = 400 \text{ ms}/(6) \times 11.14 \ \Omega = 6 \ \mu F$$

Finally, calculate the ripple using the table of Fig. 3–59. As shown, maximum ripple (lowest frequency) is about 12 mV at an engine rpm of 300 (a frequency of 1,000 Hz). This estimated 12 mV is still well below 19.5 mV (the 1/2 LSB voltage for the 8-bit ADC).

3.3.5 Ripple versus Response Time

Figures 3–60 and 3–61 show typical ripple output and response time, respectively, for our circuit. Note that the actual response time is only 100.1 ms, much faster than the 400-ms engine response time given in the example. In many applications, a compromise must be made between ripple and response time. Here are some guidelines for simplified design.

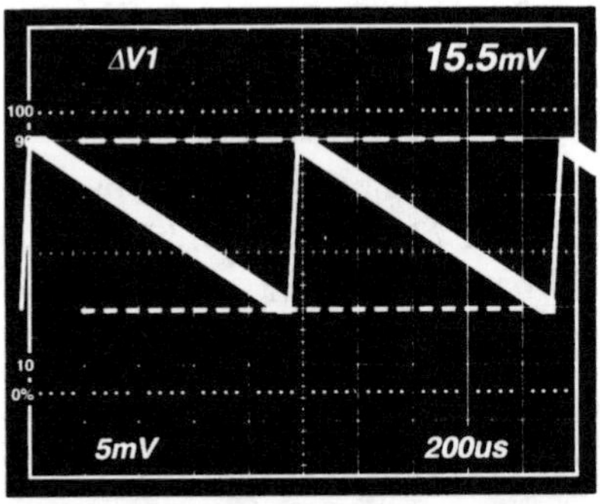

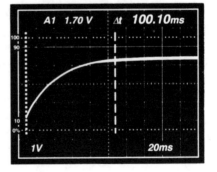

Figure 3–60.
Typical ripple output (Analog Devices, *Applications Reference Manual,* 1993, p. 23-62)

Figure 3–61.
Response to step change in frequency (Analog Devices, *Applications Reference Manual,* 1993, p. 23-62)

If response time is of primary importance, the value of C_{INT} can be lowered, but at the expense of increased ripple. On the other hand, if ripple is paramount, C_{INT} can be increased. However, this results in slower response. In the example of Section 3.3.4, it is assumed that high ripple content is the less-desirable effect, so the value of C_{INT} is kept high. In Figs. 3–60 and 3–61, the value of C_{INT} is lowered (to about 1.5 μF). This increases the ripple from about 12 mV to 15.5 mV.

3.4 Analog-to-Digital Conversion with VFC and Microprocessors

This section describes how VFCs can be combined with microprocessors to perform analog-to-digital (ADC) conversion. There are two basic methods, pulse counting and period timing. Both techniques are discussed, together with several examples (including circuit connections and software routines).

3.4.1 Pulse-Counting ADCs

Figure 3–62 shows the basic pulse-counting ADC where an AD654 VFC is used with an Intel 8051 microcomputer. Figure 3–63 shows the 8051 register. Figure 3–64 shows the 8051 pulse-count routine. Note that the 8051 is a member of the classic Intel MCS-51 family of 8-bit microcomputers. In this example, the term "8051" refers to all members of the MCS-51 family.

In this basic application, the 8051 counts the number of pulses that occur in a fixed time period. The total number of pulses counted during this period is then proportional to the input voltage of the AD654. For example, if a 1-V full-scale input produces a 100-kHz signal from the AD654, and the count period is 100 ms, then the

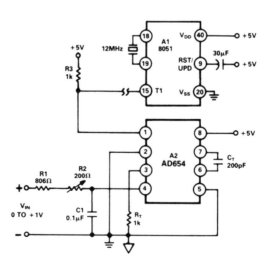

Figure 3–62.

Basic pulse-counting ADC (Analog Devices, *Applications Reference Manual*, 1993, p. 23-6)

Figure 3–63.
8051 TMOD register
(Analog Devices, *Applications Reference Manual,*
1993, p. 23-6)

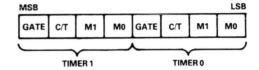

total full-scale count is 10,000. Scaling from this maximum is then used to determine the input voltage. That is, a count of 5,000 corresponds to an input voltage of 0.5 V.

The analog input to the circuit of Fig. 3–62 is a 0 to +1-V signal. The timing resistor and capacitor, R_T and C_T, are selected so that the input signal seen at pin 4 results in a 0 to 500-kHz output frequency. Resistor R2 is the full-scale adjustment. The pull-up resistor R3 ensures that the AD654 output meets the logic levels required at T1 (pin 15) of the 8051.

The 8051 has two 16-bit timer/event counters on-chip (the 8052 and 8053 have three). These counters, Timer 0 and Timer 1, can be programmed independently to operate as 16-bit time-interval or event counters. The use of Timer 0 and Timer 1 is determined by two 8-bit registers: TMOD (timer mode) and TCON (timer control). The TMOD register is shown in Fig. 3–63.

M1 and M0 bits are used to select the mode of each timer. The Mode 01 bit configures the timer as a 16-bit time-interval or event counter. The C/T is the timer or counter selector and is cleared for timer operation. In this application, Timer 0 is configured as the timer (to provide the fixed time interval) and Timer 1 is configured as the counter (to count the pulses). In the following discussion, the two timers are referred to as Timer 0 and Counter 1.

When running, Timer 0 increments at a rate equal to the external clock divided by 12. Using a 12-MHz crystal as shown in Fig. 3–62, Timer 0 increments once every microsecond.

The GATE bit is the gating control. When GATE is clear, the corresponding timer/counter is enabled whenever the corresponding TR control bit found in the TCON register is set. (The TR bit is controlled via software.) When the GATE bit is set, the corresponding timer/counter is enabled whenever the corresponding TR control bit is set, and the signal level appearing at the $\overline{\text{INT}}$ pin (pin 12 or 13 for Timer 0 or 1, respectively) is high. As a result, when a GATE bit is clear, that timer is controlled by software only. When GATE is set, the timer is controlled by a combination of software and hardware. In the circuit of Fig. 3–62, both GATE bits are clear.

Figure 3–64 lists the software routine, PLSECNT, which counts the number of falling edges that appear at T1 (the Counter 1 input) in a 50-ms window for the circuit of Fig. 3–63. After Counter 1 is cleared, the value 15539 is loaded into Timer 0. Because Timer 0 is a 16-bit timer, the maximum possible count is 65535. With the Timer 0 interrupt enabled, a count of 65536 causes a jump to the starting address (OBH) of the Timer 0 interrupt service program. By making the count start at 15539, and incrementing once every microsecond (based on a 12-MHz clock), there are

```
            ORG   00H
            AJMP  MAIN
PLSECNT     ORG   60H              ;PULSE COUNT SUBROUTINE
            MOV   TMOD, #51H       ;Put Time 0 and Count 1 in Mode 01
            MOV   TL1, #00H        ;Initialize Counter 1 Register
            MOV   TH1, #00H
            MOV   TL0, #0B3H       ;Load Time 0 With 15536 + 3.Will
            MOV   TH0, #3CH        ;Ovflw After 50ms + 3µs Delay
            SETB  PT0              ;Prioritize Time 0 Interrupt
            SETB  ET0              ;Enable Time 0 Interrupt
            SETB  EA               ;Enable Global Interrupt
            SETB  TR0              ;Start Timer
            SETB  TR1              ;Start Counter
            RET                    ;Return to Main Program

            ORG   0BH              ;TIME 0 INTERRUPT SUBROUTINE
            CLR   TR1              ;Stop Counter
            CLR   TR0              ;Stop Timer
            AJMP  COUNT

            ORG   40H
COUNT       MOV   50H,TL1          ;Move Counter Contents Into RAM
            MOV   51H,TH1
            RETI                   ;Return from Interrupt

            ORG   100H
MAIN        -     -                ;Main Program for Which PLSECNT
                                   ;Subroutine Was Written
```

Figure 3-64. 8051 pulse-count routine (Analog Devices, *Applications Reference Manual,* 1993, p. 23-6)

49,997 counts (or 49,997 ms) before jumping to the service program. The 3-µs difference from 50 ms is made up in the speed of the interrupt response. The interrupt latency ranges from 3 µs to 7 µs when using a 12-MHz crystal.

During the 50-ms count period, control resides with the main program, so the 8051 is not tied up while Counter 1 is counting for 50 ms. After the interrupt service is reached, Counter 1 and Timer 0 are stopped and the contents of Counter 1 are moved into RAM where they can be accessed at the user's convenience. Control is then returned to the main program for which the subroutine of Fig. 3–64 was written.

With a maximum frequency of 500 kHz (the limit of the AD654), and a count window of 50 ms, the maximum value of Counter 1 is 25,000. This provide resolution greater than 14 bits. Appropriate scaling from this 1-V full-scale reference point can then be performed in software.

3.4.2 Period-Timing ADCs

Figure 3–65 shows the basic period-timing ADC where an AD650 is used with an SN7474 flip-flop (FF) and an 8051. In this application, the 8051 times the period of the VFC output frequency. For example, an output frequency of 25 kHz has a period of 40 µs. If a timer that is incremented once per microsecond is gated to this sig-

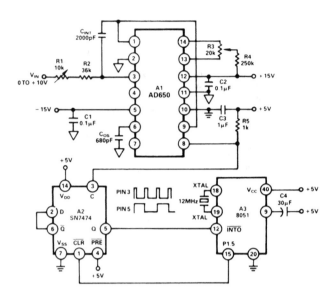

Figure 3–65. Basic period-timing ADC (Analog Devices, *Applications Reference Manual,* 1993, p. 23-7)

nal, a total count of 40 will result. An output frequency 25 Hz has a period of 4 ms. The same timer gated to this period signal then produces a total count of 4,000.

With period-timing ADCs, the count window depends on the output frequency of the VFC. This is an advantage over the pulse-counting ADC because (in most cases) the count window is shorter for period-timing than for pulse-counting. (The shorter count window is especially helpful in systems in which a number of channels are being converted.)

As an example, in the pulse-count system of Fig. 3–62, the count window is 50 ms, whether the output frequency is 50 kHz or 50 Hz. With period-timing, the count window is the inverse of the output frequency. A 50-kHz signal has a count window of 20 μs, and a 50 Hz signal has a window of 20 ms. Not until the output frequency reaches 20 Hz does the period-timing count window equal the 50-ms pulse-count window.

The analog input to the circuit of Fig. 3–65 is a 0 to +10-V signal that results in a 0 to 50-kHz output. Because the AD650 output is made up of pulses, the SN7474 D-type FF is used to convert these pulses into square waves. The SN7474 pin-3 and pin-5 waveforms in Fig. 3–65 show that the width of either the high-level or low-level output appearing at pin 5 is the same as one period of the AD650 output frequency. Note that when the pin of the SN7474 is held low, pin 5 is also held low.

The INTO pin (pin 12) of the 8051 is the Timer-0 gate pin. When the GATE bit is set in the TMOD register (Fig. 3–63), Timer 0 runs only when INTO at pin 12 is high, and the TR0 bit in the TCON register is set via software. Connecting the Q output of the SN7474 to the INTO pin ensures that the timer runs for one period of the AD650 frequency.

It is possible that the TR0 bit (in software) might be set in the middle of a high-level edge at the INTO pin on the 8051. In this case, Timer 0 runs for a fraction of a period, rather than one full period. This problem is overcome by tying the 8051 Port 1 bit 5 (P1.5) to the SN7474 CLR pin. When CLR is low, and PRE is high, Q is low. When CLR and PRE are both high, Q changes state on every positive edge appearing at the clock (C) pin. Setting P1.5 low, then setting TR0 (in software), and then setting P1.5 high ensures that Timer 0 runs for one full period.

Figure 3–66 lists the software routine, PCNT, which increments Timer 0 once per microsecond for one AD650 output-frequency period. Note that there are two interrupt service programs, one for INTO and one for Timer 0.

```
           ORG    00H
           AJMP   MAIN
PCNT       ORG    90H            ;PERIOD COUNTER SUBROUTINE
           MOV    TMOD, #05H     ;Put Time 0 in Mode 1, Enable INTO Pin
           CLR    P1.5           ;Initially Set INTO Pin Low
           SETB   IT0            ;Specify Edge Triggered Interrupt
           MOV    TL0, #00H      ;Initialize Timer
           MOV    TH0, #00H
           SETB   EX0            ;Enable INTO Interrupt
           SETB   ET0            ;Enable Timer 0 Interrupt
           SETB   EA             ;Enable All Interrupts
           SETB   TR0            ;Start Timer
           SETB   P1.5           ;Enable Gate INTO Pin
           RET                   ;Return to Main Program

           ORG    03H            ;INTO Subroutine Service Program
           CLR    TR0            ;Stop Timer
           CLR    EA             ;Disable Interrupts
           AJMP   COUNT          ;Jump to Count

           ORG    0BH            ;TIMER 0 SUBROUTINE
                                 SERVICE PROGRAM
           CLR    TR0            ;Stop Timer
           CLR    EA             ;Disable Interrupts
           AJMP   OFLW           ;Jump to OFLW

           ORG    40H
OFLW       MOV    60H, #FF       ;Load RAM With Overflow
           MOV    61H, #FF       ;Value
           CLR    P1.5           ;Set INTO Pin Low
           RETI                  ;Return From Subroutine

           ORG    50H
COUNT      MOV    60H,TH0        ;Load RAM with Counter Contents
           MOV    61H,TL0
           CLR    P1.5           ;Set INTO Pin Low
           RETI                  ;Return From Subroutine

           ORG    100H
MAIN       -      -              ;Main Program for Which
                                 ;Subroutine Was Written
```

Figure 3–66. Period-timing routing (Analog Devices, *Applications Reference Manual,* 1993, p. 23-7)

The $\overline{\text{INT0}}$ service program is accessed after a negative edge appears at the $\overline{\text{INT0}}$ pin, indicating the end of one period. The timer is then stopped and the timer contents are loaded into RAM. The user can then access the contents as required.

The Timer-0 service program limits the count window to about 65.5 ms, and is accessed when the contents of Timer 0 reach 65536. This occurs when the AD650 input voltage is about 3.05 mV, or the output frequency is about 15.26 Hz. The Timer-0 service program then loads the overflow value 65535 into RAM.

After both interrupt subroutines, and after initialization of the PCNT subroutine, control returns to the main program for which the subroutine was written. As a result, the 8051 is not tied up during the period-timing interval. Resistor R3 sets the zero-offset (adjust for zero output with the input at zero or grounded). Resistor R1 is the full-scale adjustment (adjust for 50 kHz with +10 V at the input).

Jitter, or the range of variation in the output-frequency period, can be a problem in period-timing ADCs. This variation in period might result in a variation in the number of pulses counted from one period to the next. The magnitude of such jitter error can be reduced in software by taking the average of a number of period counts, and using the average value for calculations.

3.4.3 ADC with 16-Bit Resolution

Figure 3–67 shows a complete ADC system with 16-bit resolution, including display LEDs, using an AD651 as the VFC. The AD651 has a 2-MHz full-scale output in this application, but requires an external clock (4 MHz TTL) as shown. The system also includes an Intersil 7208 single-chip counter-decoder-LED driver, two 4020B binary counters, and a 74LS221 dual monostable multivibrator. Note that this system does not require a microprocessor, so a software routine is not needed. Instead, the 7208 provides decoding and drive for the LEDs, as well as the count function.

The analog input to the circuit of Fig. 3–67 is a 0 to +10-V signal that results in a 0 to 2-MHz output. Resistor R8 sets the zero-offset (adjust for zero output on the LEDs with the input at zero or grounded). Resistor R10 is the full-scale adjustment. Pull-up resistor R1 feeds the AD651 output frequency directly into the counter-input pin of the 7208.

The 4020Bs are 14-stage binary ripple counters that have clock and master reset inputs (pins 10 and 11), and buffered outputs from the first stage, and the last eleven stages. The 4020Bs generate the count window for the 7208. A signal of 15.2588-Hz is produced by dividing the 4-MHz TTL clock by 2^{18} (two divide-by-2^9 stages cascaded). The 7208 counts the negative edges of the AD651 output pulses when pin 13 (count enable) is low. With a 15.2588-Hz signal, the count window is 32.768 ms.

Figure 3–68 shows the relationship between the AD651 clock frequency and the gate time for various degrees of resolution. In the circuit of Fig. 3–67, 16 bits of resolution are required, using a 4-MHz clock, so the conversion or gate time is 32.77 ms.

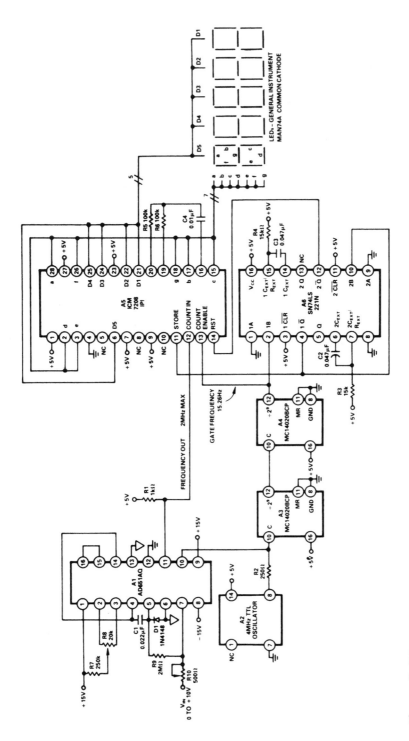

Figure 3-67.
ADC system with 16-bit resolution (Analog Devices, *Applications Reference Manual*, 1993, p. 23-8)

Resolution	N	Clock	Conversion or Gate Time	Typ Lin	Comments
12 Bits	4096	81.92kHz	100ms	0.002%	50,60,400HzNMR
12 Bits	4096	2MHz	4.096ms	0.01%	
12 Bits	4096	4MHz	2.048ms	0.02%	
4 Digits	10000	200kHz	100ms	0.002%	50,60,400HzNMR
14 Bits	16384	327.68kHz	100ms	0.002%	50,60,400HzNMR
14 Bits	16384	1.966MHz	16.66ms	0.01%	60HzNMR
14 Bits	16384	1.638MHz	20ms	0.01%	50HzNMR
4 1/2 Digits	20000	400kHz	100ms	0.002%	50,60,400HzNMR
16 Bits	65536	655.36kHz	200ms	0.002%	50,60,400HzNMR
16 Bits	65536	4MHz	32.77ms	0.02%	

Figure 3–68. Relationship between clock frequency and gate time for various degrees of resolution (Analog Devices, *Applications Reference Manual,* 1993, p. 23-9)

Also note the relationship among resolution, clock frequency, and typical linearity. For example, as shown in Fig. 3–68, 12-bit resolution with a 2-MHz clock results in 0.01% linearity. If the clock is increased to 4 MHz, the linearity is 0.2%. For simplified design, the fixed gate interval should be generated using a multiple of the clock input. This eliminates clock errors (jitter, drift with time or temperature, and so forth) because it is the ratio of the clock and output frequencies that is being measured.

Figure 3–69 shows the 7208 timing diagrams. After the count window on the 7208 is closed (that is, after the count-enable gate signal has gone high), the data bits are latched and then decoded to drive the LEDs. After the data bits are latched, the counter must then be cleared to get an accurate count during the next count window. This is done with store and reset pulses from the 74LS221. (Note that the 15.26-Hz count-enable signal is also applied to the 74LS221.)

Although the 7208 only requires pulse widths greater than 50 μs, the values of R_3/C_2 and C_2/R_4 are selected to provide a pulse width of about 500 μs. The 7208

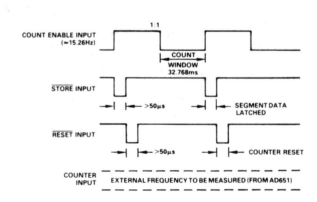

Figure 3–69.
7208 timing diagrams
(Analog Devices, *Applications Reference Manual,* 1993, p. 23-9)

automatically decodes data and drives the LEDs. The values of R_5, R_6, and C_4 are selected (using the manufacturer's guidelines) to provide a multiplex rate between 50 and 200 Hz.

3.4.4 Interference Signals in ADCs

A common problem in ADCs is an interfering signal that couples into the analog signal being converted. For example, undesired coupling from the power line often appears as a sine wave riding on top of the DC level to be converted, resulting in an erroneous digital output. Because the frequency of this interfering sine wave is known (50 or 60 Hz), the errors caused by the pickup can be integrated out with a gating time equal to multiple of the interfering sine-wave period.

Figure 3–70 shows the basic block diagram of a phase-locked loop (PLL) used to gate off an interfering signal in an ADC. Figure 3–71 shows the practical circuit. A replica of the interfering signal is picked off from any convenient point (say, a transformer) and fed into the PLL that provides two output signals: a high-frequency clock (at a high harmonic of the interference) and a gating clock (at a lower harmonic of the interference). By using f_{OUT} as the AD651 clock source, and f_{IN} from the divide-by-N counter as the count-window source, a resolution of one part in 1/2 N is achieved (where N is the "divide by N" of the counter). If the count window is level-triggered rather than edge-triggered (as with the 7208), then the resolution is 1/4 N.

The MC4044 in Fig. 3–71 contains both the phase detector and the amplifier/filter shown in Fig. 3–70. In this case, the interfering signal is 60 Hz and is converted into a TTL level to feed into the MC4044. (Components R_1, R_2, and C_1 are selected to allow for a 50-Hz or 60-Hz interference signal.)

The error voltage from the MC4044 is fed into the AD654 that is configured in a 0 to +1-V input, with a 0 to 500-kHz output. The 74LS393 counters are connected to provide a divide-by-N of 8192, so the output of the AD654 is 491520 Hz. This signal is used as the clock for the AD651. The output of the second counter is 60 Hz, and is fed back to the MC4044, in addition to being used as the frequency-counter gate signal.

Resolution can be increased using pins 10, 9, or 8 of A4 as the frequency-counter gate signal. However, pin 11 of A4 must be fed back to the MC4044. Connecting from pin 1 of A3 to pin 11 of A4 produces N = 8192, or 2^{13}. This supplies 12-bit resolution. Using pins 10, 9, or 8 to provide the gate frequencies will produce resolutions of 13, 14, or 15 bits, respectively.

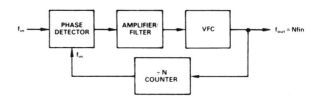

Figure 3–70.

Basic phase-locked loop (Analog Devices, *Applications Reference Manual,* 1993, p. 23-9)

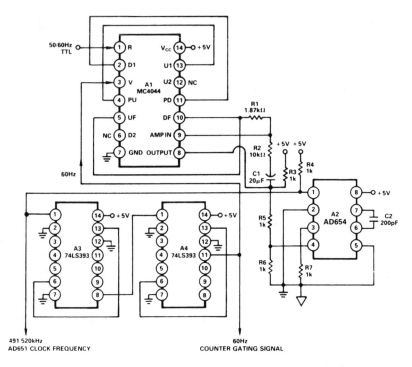

Figure 3–71. Circuit for gating off a known interference signal (Analog Devices, *Applications Reference Manual,* 1993, p. 23-10)

3.4.5 ADC Using a Microprocessor Without an On-Chip Timer

Figure 3–72 shows an ADC using an AD650 as the VFC and an MC6801 microprocessor. Although the MC6801 has an on-chip timer, it is assumed that the timer function is used for some other purpose and is not available for the ADC function. Using additional ICs, it is still possible to count the output pulses of the VFC for a fixed period of time and store the count in the microprocessor RAM. Figure 3–73 shows the pulse-count routine.

The MC6801 is an 8-bit device with 2,048 bytes of ROM, 128 bytes of RAM, a serial communication interface, and a three-function programmable timer. The AD650 is connected for a 0 to +10-V input with a 0 to 1-MHz output. The additional ICs used to count the output pulses of the AD650 include two 74590 8-bit binary counters with output registers, one 4020B 14-stage binary counter, one 7474 FF with preset and clear, and one inverter of a 74LS504.

The 4020B and 7474 provide the timing that tells the counters when to start and stop counting. The input to the 4020B is tied to the E pin (pin 40) of the MC6801. The signal at the E pin is the MC6801 external-crystal frequency divided by four or,

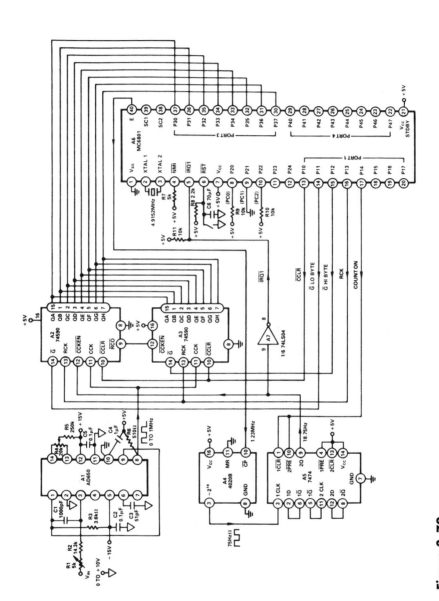

Figure 3-72.
ADC system with microprocessor (Analog Devices, *Applications Reference Manual*, 1993, p. 23-11)

			ORG	0100		COUNT ROUTINE
8E	00C0	BEGIN	LDS	#$C0		Set Stack Pointer
0F			SEI			Disable Interrupts
86	06		LDAA	#$06		Clear Counter
97	02		STAA	$02		
86	17		LDAA	#$17		Turn On Counter
97	02		STAA	$02		
86	2F		LDAA	#$2F		Insert Delay >13.33ms To
C6	7F	AGN	LDBB	#$7F		Prevent Interrupt Error
5A		CNT	DECB			
2E	FD		BGT	CNT		
4A			DECA			
2E	F8		BGT	AGN		
0E			CLI			Allow Interrupt
39			RTS			Return From Subroutine
			ORG	FFF8		
0080			AIRQ1	FDB	LDCNT	Define Interrupt Routine
						Start Point
			ORG	0080		INTERRUPT ROUTINE
0F		LDCNT	SEI			Disable Interrupt
86	07		LDAA	#$07		Turn Off Counter
97	02		STAA	$02		
86	0F		LDAA	#$0F		Latch Data In Counter
97	02		STAA	$02		
86	0D		LDAA	#$0D		Output Counter Lo Byte
97	02		STAA	$02		
96	06		LDAA	$06		Read Byte From Port 3
97	A7		STAA	$A7		Store in Location 00A7
86	0B		LDAA	#$0B		Output Counter Hi Byte
97	02		STAA	$02		
96	06		LDAA	$06		Read Byte From Port 3
97	A6		STAA	$A6		Store In Location 00A6
86	07		LDAA	#$07		Turn Off Counter
97	02		STAA	$02		
86	10		LDAA	#$10		Set The Interrupt Bit On The
9A	BA		ORAA	$BA		Stacked Condition Code Register
97	BA		STAA	$BA		
3B			RTI			Return From Interrupt

Figure 3–73. Pulse-count routine for ADC/microprocessor (Analog Devices, *Applications Reference Manual,* 1993, p. 23-11)

in this case, 1.2288 MHz. The 4020B divides the 1.2288-MHz signal by 2^{14}, which results in a 75-Hz signal being fed into pin 3 of the 7474.

The 7474 further divides the signal by four to 18.75 Hz. This signal appears at pin 9, depending on the signal level at pins 1 and 10. If pins 1 and 10 are low, pin 9 will be held at a TTL high. If pins 1 and 10 are high, the 18.75-Hz square wave will appear at pin 9. When pin 9 is high, the 74590 counters are disabled and no counting occurs. If the output of pin 9 is a square wave, the counters are enabled during the low level of the period and will thus count for 26.67 ms. Note that it is bit 4 of port 1 (P14 of the MC6801) that controls whether the counters are enabled. (If bit 4 of port 1 is low, no count occurs.)

Pin 9 of the 7474 is connected to the external interrupt-request ($\overline{\text{IRQ1}}$) pin of the MC6801 through an inverter. This produces two results when pin 9 changes from low to high. First, the counters are disabled because $\overline{\text{CCKEN}}$ of A2 is tied to pin 9. Second, the low-to-high transition is inverted and generates an interrupt request on the $\overline{\text{IRQ1}}$ line. The MC6801 then clears P14 to prevent another count from occurring before the 74590 counter values are read.

All of the counting events are controlled by the signal levels of the different bits of port 1. By writing different values to port 1, the 74590 counters are cleared, enabled, disabled, latched, and read. The values that control these functions are shown in Fig. 3–74.

The routine shown in Fig. 3–73 sets the stack pointer, disables the interrupts, clears the 74590 counters, and then enables the interrupts after a delay period. When the interrupt request is triggered by a low-to-high transition at pin 9 of the 7474, the program counter jumps to the interrupt routine.

The interrupt routine disables any further interrupts, turns off the 74590 counters, and reads the low and high bytes, storing them in locations 00A7 and 00A6, respectively. The interrupt bit on the stacked condition-code register is then set, and the program counter returns from the interrupt routine.

The delay time before enabling the MC6801 interrupt is shown in Fig. 3–75. That is, Fig. 3–75 shows the waveforms that appear near the time that the 74590 counters are instructed to start counting by setting pins 1 and 10 high. As discussed, pin 9 outputs the 18.75-Hz square wave only when pins 1 and 10 are held high. Pin 9 does not change state after pins 1 and 10 go high until the next positive edge occurs at pin 3. Figure 3–75 shows the worst case (where pins 1 and 10 go high just a few nanoseconds after pin 3 sees a positive-going edge). Pin 9 will not change state for about another 13.3 ms.

The $\overline{\text{IRFQ1}}$ signal from pin 9 is fed through inverter A7, so the signal level is low during the 13.33-ms wait interval. Because $\overline{\text{IRFQ1}}$ is level sensitive, allowing an interrupt during the wait state would cause a jump to the interrupt routine before the 26.67-ms count window is open. By inserting a delay greater than 13.33 ms before allowing the interrupt, the interrupt is enabled during the count window (when pin 9 is low, and the $\overline{\text{IRFQ1}}$ pin is high). This ensures that the interrupt routine is accessed after the 26.67-ms count window closes.

After executing the CLI instruction in software (Fig. 3–73), bit 4 of the condition-code register (CCR) is cleared. This enables the $\overline{\text{IRQ1}}$ interrupt. When an inter-

Figure 3–74.
Port 1 event-control values (Analog Devices, *Applications Reference Manual,* 1993, p. 23-11)

Port 1 Configuration						
Event	P4	P3	P2	P1	P0	Hex
Clear Counters	0	0	1	1	0	06
Enable Counters	1	0	1	1	1	17
Disable Counters	0	0	1	1	1	07
Latch Data	0	1	1	1	1	0F
Output Hi Byte	0	1	1	0	1	0D
Output Lo Byte	0	1	0	1	1	0B

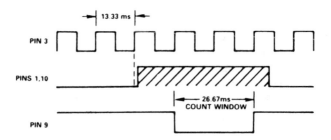

Figure 3–75. SN7474 waveforms (Analog Devices, *Applications Reference Manual,* 1993, p. 23-12)

rupt request is detected (a low level at the $\overline{\text{IRQ1}}$ pin), the CCR is pushed onto the stack in its present state. Because the stack pointer is set at location 00C0, the location of the CCR is 7 bytes down, or at location 00BA. By OR'ing 10H with the contents of location 00BA, the interrupt is disabled.

With the interrupt disabled at this point, the program counter is prevented from jumping into the interrupt routine again immediately after coming out of the routine, even though the level sensed at the $\overline{\text{IRQ1}}$ pin is low. (If the bit were not set, the program counter would jump out of the interrupt routine, see that the interrupt-enable bit is clear and that the level at the $\overline{\text{IRFQ1}}$ pin is low, and jump back into the interrupt routine again.)

CHAPTER **4**

Simplified Design with Exar VFCs

This chapter is devoted to simplified-design approaches for Exar Corporation VFCs (the XR-4151). All of the general design information in Chapter 1 applies to the examples in this chapter. However, each voltage-frequency IC has special design requirements. The circuits in this chapter can be used immediately the way they are, or by altering component values, as a basis for simplified design of similar voltage-frequency conversion applications.

4.1 Description of XR-4151

Figure 4–1 shows the basic block diagram of the XR-4151. Figure 4–2 shows the electrical characteristics of most concern for simplified design. The absolute maximum ratings for the XR-4151 are 22 V for power supply, 20 mA for output sink current, 500 mW of internal power dissipation, –0.2 V to +V_{CC} for input voltage, and continuous operation with an output short circuit to ground.

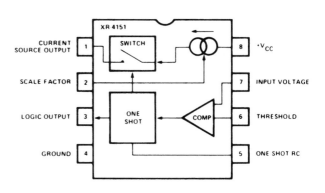

Figure 4–1.
Basic block diagram of XR-4151 (*EXAR Databook,* 1992, p. 2-328)

129

ELECTRICAL CHARACTERISTICS
Test Conditions: (V_{CC} = 15V, T_A = +25°C, unless otherwise specified)

PARAMETERS	XR-4151CP		XR-4151P		TYP	UNITS	CONDITIONS
	MIN	MAX	MIN	MAX			
Supply Current	2.0	6.0	2.0	6.0	3.5	mA	8V < V_{CC} < 15V
	2.0	7.5	2.0	7.5	4.5	mA	15V < V_{CC} < 22V
Conversion Accuracy							
Scale Factor	0.90	1.10	0.92	1.08	1.00	kHz/V	Circuit of Fig. 3, V_I = 10V, R_S = 14.0k
Drift with Temperature	–	–	–	–	±100	ppm/°C	Circuit of Fig. 3, V_I = 10V
Drift with V_{CC}	–	–	–0.9	0.9	0.2	%/V	Circuit Fig. 3, V_I = 1.0V
							8V < V_{CC} < 18V
Input Comparator							
Offset Voltage	–	10	–	10	5	mV	
Offset Current	–	±100	–	±100	±50	nA	
Input Bias Current	–	–300	–	–300	–100	nA	
Common Mode Range (Note 1)	0	V_{CC}–3.0	0	V_{CC}–3.0	0 to V_{CC} –2	V	
One-Shot							
Threshold Voltage, Pin 5	0.63	0.70	0.63	0.70	.667	xV_{CC}	Pin 5.1 = 2.2mA
Input Bias Current, Pin 5	–	–500	–	.500	–100	nA	
Reset V_{SAT}	–	50.0	–	50.0	0.15	V	
Current Source							
Output Current	–	–	–	–	138.7	µA	Pin 1, V = 0, RS = 14.0kΩ
Change with Voltage	–	2.5	–	2.5	1.0	µA	Pin 1, V=0V to V = 10V
Off Leakage	–	50.0	–	50.0	0.15	nA	Pin 1, V = 0V
Reference Voltage	1.70	2.08	1.70	2.08	1.9	V	Pin 2
Logic Output							
V_{SAT}	–	0.50	–	0.50	0.15	V	Pin 3, 1 = 3.0mA
V_{SAT}	–	0.30	–	0.30	0.10	V	Pin 3, 1 = 3.0mA
Off Leakage	–	1.0	–	1.0	.1	µA	

Note 1: Input Common Mode Range includes ground.

Figure 4-2.
XR-4151 electrical characteristics (*EXAR Databook*, 1992, p. 2-329)

4.2 Basic VFC Operation

Figure 4–3 shows the connections for stand-alone VFC operation with a single power supply. As shown, the XR-4151 contains a voltage comparator, a one-shot, and a precision switched current source. The voltage comparator compares a positive input voltage applied at pin 7 to the voltage at pin 6. If the input (pin 7) voltage is higher, the comparator fires the one-shot.

The one-shot output is connected to both the logic output and the switched current source. During the one-shot period (T), the logic output goes low and the current source turns on with current (I). At the end of the one-shot period, the logic output goes high and the current source shuts off. At this time, the current source has injected an amount of charge $Q = I_OT$ into the network R_BC_B.

If the charge has not increased the voltage V_B such that V_B is greater than V_I, the comparator again fires the one-shot and the current source injects another charge (Q) into the R_BC_B network. This process continues until V_B is greater than V_I. This completes one cycle. The VFC will now run in a steady-state mode.

When running, the current source dumps charges into capacitor C_B at a rate fast enough to keep V_B equal to (or greater than) V_I. Because the discharge rate of capacitor C_B is proportional to V_B/R_B, the frequency at which the system runs is proportional to the input voltage.

4.2.1 Basic VFC Design Example

Using the values shown in Fig. 4–1, the input voltage range is from 0 to +10 V, and the output frequency is from 0 to 10 kHz. The full-scale output frequency is adjusted by R_S. Although this circuit is simple, there are certain design limitations.

First, the linearity error is typically 1%. Next, a frequency offset is introduced by the input comparator offset voltage. Response time for the circuit is limited by the passive integration network R_BC_B. (Note that C_B should have a high-stability dielectric, such as mica, polystyrene, or polyester.) Typical response time to a step-change

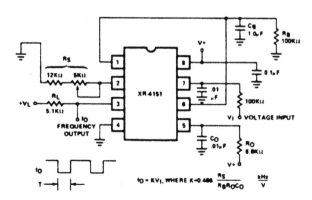

Figure 4–3.
XR-4151 connected as a basic VFC (*EXAR Databook,* 1992, p. 2-330)

input from 0 to +10 V is 135 ms. For applications that require fast response time and high accuracy, use the precision circuit described in Section 4.3.

4.3 Precision VFC

Figure 4–4 shows the XR-4151 used with an op-amp integrator to provide a VFC with typical linearity of 0.05% over the range of 0 to –10 V. The offset is adjustable to zero. Unlike some VFC designs that lose linearity below 10 mV, this circuit retains linearity over the full range of input voltage, all the way to 0 V.

To calibrate the circuit, apply –10 V and adjust R5 for an output frequency of 10 kHz. Then apply –10 mV and adjust the offset pot for an output frequency of 10 Hz. Repeat these adjustments as necessary.

The op-amp integrator improves linearity of this circuit over that of Fig. 4–3 by holding the source output (pin 1) at a constant 0 V. As a result, because of the current-source output conductance, the linearity error is eliminated. The diode connected around the op-amp input/output prevents the voltage at pin 7 of the XR-4151 from going below zero. (Use a low-leakage diode because any leakage will degrade the accuracy.)

This circuit can be operated from a single positive supply if an XR-3403 ground-sensing op amp is used for the integrator. In this case, the diode can be omitted. Note that even though the circuit will operate from a single supply, the input voltage must be negative. For operation above 10 kHz, bypass pin 6 of the XR-4151 with 0.01 µF.

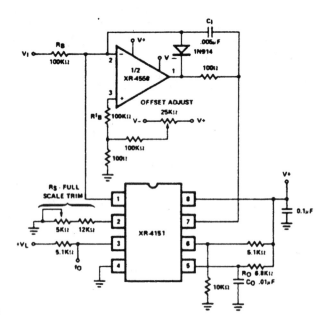

Figure 4–4.
XR-4151 connected as a
precision VFC (*EXAR
Databook,* 1992, p. 2-330)

4.4 Basic FVC Design Example

Figure 4–5 shows the XR-4151 connected to provide basic FVC operation, using a single supply. With no signal applied, the resistor bias networks tied to pins 6 and 7 hold the input comparator in the off state. A negative-going pulse applied to pin 6 (or positive pulse at pin 7) causes the comparator to fire the one-shot.

For proper operation, the pulse width must be less than the period of the one-shot, $T = 1.1 \, R_O \, S_O$. For a 5-Vp-p square-wave input, the differentiator network formed by the input coupling capacitor and the resistor bias provides pulses that correctly trigger the one-shot. An external voltage comparator can be used to "square up" sine-wave signals before they are applied to the XR-4151.

The component values for the input-signal differentiator and bias network can be altered to accommodate square waves with different amplitudes and frequencies. The passive integrator network $R_B C_B$ filters the current pulses from the pin-1 output. For less output ripple, increase the value of C_B.

4.5 Precision FVC Design Example

Figure 4–6 shows the XR-4151 used with an op-amp integrator to provide an FVC with increased accuracy and linearity. The full-scale adjust is trimmed for –10-V output with 10-kHz input. The offset adjust is trimmed to give a –10-mV output with a 10-Hz input.

There is a tradeoff between response time and output ripple, as determined by the value of C_I. If C_I is 0.1 μF, the ripple is about 100 mV. The response-time constant t_R is $R_B C_I$. For an R_B of 100 k, and a C_I of 0.1 μF, t_R is about 10 ms.

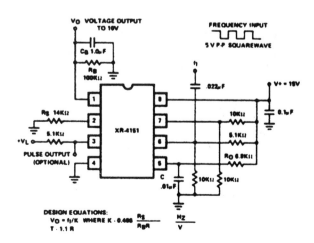

Figure 4–5.
XR-4151 connected as a basic FVC (*EXAR Databook*, 1992, p. 2-331)

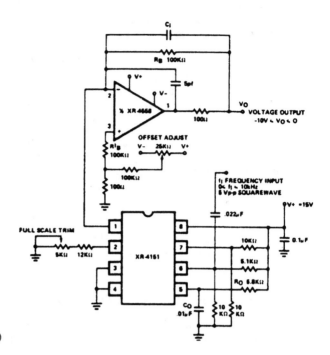

Figure 4–6.
XR-4151 connected as a
precision FVC (*EXAR
Databook,* 1992, p. 2-331)

4.6 XR-4151 Design Considerations

The following considerations should be made when using the XR-4151 for all
the circuits:

- The voltage applied to comparator input pins 6 and 7 should not be allowed
  to go below ground by more than 0.3 V.
- Pins 3 and 5 are open-collector outputs. Shorts between these pins and
  +VCC can cause overheating and eventual destruction.
- Reference-voltage terminal pin 2 is connected to the emitter of an NPN transistor, and is held at about 1.9 V. This terminal should be protected from ac-
  cidental shorts to ground or supply voltages. Permanent damage might occur
  if the current in pin 2 exceeds 5 mA.
- Avoid stray coupling between pins 5 and 7. Such coupling might cause false
  triggering. For the circuit of Fig. 4–3, bypass pin 7 to ground with at least
  0.01 μF. This is necessary for operation above 10 kHz.
- For the best stability over time, use a silver mica, polystyrene, or polyester
  dielectric for C_B.

4.7 Programming the XR-4151

The XR-4151 can be programmed to operate with a full-scale frequency anywhere from 1 Hz to 100 kHz. In the case of the VFC operation, nearly any full-scale input voltage from 1.0 V and up can be tolerated if proper scaling is used. The following procedures can be used for any desired full-scale frequency.

1. Set R_S to 14 k, or use a 12-k resistor and a 5-k pot, as shown in the schematics.
2. Let the one-shot period $T = 1.1 R_O C_O$ be equal to 0.75 (1/fo), where fo is the desired full-scale frequency. For best performance, make 6.8k > R_O > 680 k and 0.001 μF < C_O < 1.0 μF.
3. For the circuit of Fig. 4–3, make $C_B = 10^{-2}(1/fo)$ farads. Smaller values of CB will produce a faster response time, but will increase the frequency offset and nonlinearity.
4. For the circuit of Fig. 4–4, make $C_I = 5 \times 10^{-5}$ (1/fo) farads. The op-amp integrator must have a slew rate of at least $135 \times 10^{-6}(1/C_I)$ volts per second, where the value of C_I is in farads.
5. For the circuit of Fig. 4–4, keep the values of C_B and R_B as shown, and use an input attenuator to give the desired full-scale input voltage.
6. For the circuit of Fig. 4–6, let $R_B = V_{IO}/100$ μA, where V_{IO} is the full-scale input voltage. As an alternate, the op-amp inverting input (summing node) (pin 1 of XR-4151) can be used as a current input, with the full-scale input current $I_{IO} = -100$ μA.

   For the circuits of Figs. 4–5 and 4–6, pick the value of C_B or C_I to give the optimum tradeoff between the response time and output ripple, for the particular application.

4.8 VFC Design Example (100 kHz, –10 V)

Using the circuit of Fig. 4–4, design a VFC with a full-scale frequency of 100 kHz and a full-scale input voltage of –10 V.

1. Let R_S be 14 k (or use the 5-k pot and a 12-k fixed resistor as shown).
2. Find the one-shot period T using the full-scale frequency, or $0.75 \times (1/10^5)$s = 7.5 μs.
3. For simplified design, use the lowest recommended value for R_O (6800) as given in step 2 of Section 4.7. With R_O at 6800, find the value of C_O. $1.1 \times 6,800 = 7480$ (about 7,500). 7.5 μs divided by 7,500 = 0.001 μF (the nearest practical value).
4. Find the value of C_I, using $C_I = 5 \times 10^{-5}$ (1/10^5) F = 500 pF.
5. Find the required minimum op-amp slew rate using 135×10^{-6} (1/500 pF) = 0.27 V/μs.
6. Find the value of R_B, using $R_B = 10$ V/100 μA = 100 k.

4.9 VFC Design Example (1 Hz, 10 V)

Using the circuit of Fig. 4–4, design a VFC with a full-scale frequency of 1 Hz and a full-scale input voltage of 10 V.

1. Let R_S be 14 k (or use the 5-k pot and a 12-k fixed resistor as shown).
2. Find the one-shot period T using the full-scale frequency, or $0.75 \times (1/1) =$ 0.75 s.
3. For simplified design, use the highest recommended value for R_O (680,000) as given in step 2 of Section 4.7. With R_O at 680,000, find the value of C_O. $1.1 \times 680,000 = 748,000$ (about 750,000). 0.75 s divided by 750,000 = 1.0 µF (the nearest practical value).
4. Find the value of C_I, using $C_I = 5 \times 10^{-5}$ (1/1) F = 50 µF.
5. Find the required minimum op-amp slew rate using 135×10^{-6} (1/50 µF) = 2.7 µV/µs.
6. Find the value of R_B, using $R_B = 10$ V/100 µA = 100 k.

4.10 FVC Design Example

Using the circuit of Fig. 4–6, design an FVC with a full-scale input frequency of 83.3 Hz and a 9-V supply. The output voltage must reach at least 0.63 of the final value in 200 ms. Determine the output ripple with the selected design values.

1. Let R_B be 14 k (or use the 5-k pot and a 12-k fixed resistor as shown).
2. Find the one-shot period T using the full-scale frequency, or $0.75 \times (1/83.3)$s = 9 ms.
3. For simplified design, let R_O be 82 k (9 ms divided by 1.1) and find the value of C_O. $1.1 \times 82,000 = 90,200$ (about 90,000). 9 ms divided by 90,000 = 0.1 µF.
4. Let the full-scale output voltage (at pin 6) equal 5 V. This is well below the 9-V supply. With a 5-V output, the value of R_B is 5 V/100 µA = 50 k.
5. With an R_B of 50 k, and a desired 200-ms time constant (t_R), the value of C_I is: $(200 \times 10^{-3}) / (50 \times 10^3) = 4$ µF.
6. With a C_I of 4 µF, find the worst-case ripple, using ripple voltage $V_R = (9$ ms $\times 135$ µA) / (4 µF) = 300 mV. Keep in mind that there is a tradeoff between response time and output ripple, determined by the value of C_I (and the value of C_B) as discussed in Section 4.7.

CHAPTER 5

Simplified Design with National Semiconductor VFCs and FVCs

This chapter is devoted to simplified-design approaches for National Semiconductor VFCs and FVCs. All of the general design information in Chapter 1 applies to the examples in this chapter. However, each voltage-frequency IC has special design requirements. The circuits in this chapter can be used immediately the way they are, or by altering component values, as a basis for simplified design of similar voltage-frequency conversion applications.

5.1 FVC Applications

This section discusses applications for the National Semiconductor LM2907 and LM2917. These FVC ICs operate from a single supply, and provide basic FVC operation with a minimum of external components. The ICs include both input and output circuits, as well as the basic FVC components, in a single package. Before we discuss specific applications, we will describe operation of the ICs.

5.1.1 Basic Operation of LM2907/LM2917

Figure 5–1 shows the basic circuit functions of the LM2907/LM2917 ICs in block form. Figure 5–2 shows the basic FVC connection, including typical external components. As shown, both versions of the IC include an input amplifier with built-in hysteresis, a charge-pump FVC, and an uncommitted output transistor. The LM2917 includes an active zener regulator, not available in the LM2907. Both versions are available in 14-pin and 8-pin DIP molded packages. Other packages are available on special order.

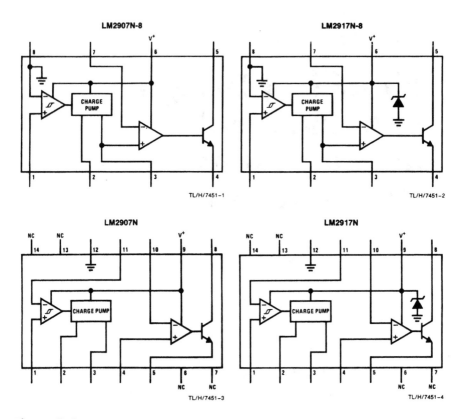

Figure 5-1. Basic circuit functions of LM2907/LM2917 (National Semiconductor, *Linear Applications Handbook*, 1994, p. 345)

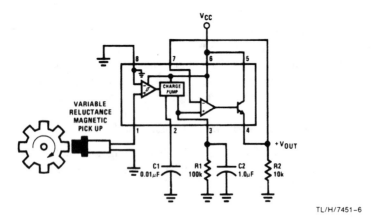

Figure 5-2. Basic FVC connections (National Semiconductor, *Linear Applications Handbook*, 1994, p. 348)

5.1.2 Basic FVC Operation

As shown in Fig. 5–2, a frequency signal (from a tachometer) is applied to the input of the charge pump at pin 1. The voltage appearing at pin 2 swings between two values, which are approximately $1/4 (V_{CC}) - V_{BE}$ and $3/4 (V_{CC}) - V_{BE}$. The voltage at pin 3 has a value equal to $V_{CC} \times f_{IN} \times C1 \times R1 \times K$, where K is the gain constant (normally 1.0).

The emitter output (pin 4) is connected to the inverting input of the op amp so that pin 4 follows pin 3, and provides a low-impedance output voltage proportional to input frequency. The linearity of this voltage is typically better than 0.3% of full scale.

5.1.3 Selecting External Components

Use the following guidelines for selecting the basic external components shown in Fig. 5–2.

Because C1 also provides internal compensation for the charge-pump circuits, the minimum value for C1 is 100 pF. Smaller values can cause an error current on R1, especially at low temperatures.

Three conditions must be met when selecting resistor R1 (at pin 3). First, the output current at pin 3 is internally fixed. (The data sheet shows a minimum of 150 mA.) The value of R1 is determined by the fixed current at pin 3, and the desired full-scale output voltage, using the equation:

$$R1 \geq V3MAX/I3MIN$$

where V3MAX is the full-scale output voltage required and I3MIN is determined from the data sheet (150 mA).

Second, if R1 is too large, it can become a significant fraction of the output impedance at pin 3, and can degrade linearity.

Third, the value of capacitor C2 is affected by the R1 value. In turn, the amount of ripple is affected by the value of C2.

An expression that describes the ripple content on pin 3 for a single R1-C2 combination is:

$$V_{RIPPLE} = V_{CC} / 2 \times C1/C2 (1 - V_{CC} \times f_{IN} \times C_1 / I_2) \text{ p-p}$$

Although it appears that R1 can be chosen without considering ripple (R1 is not in the ripple equation), the response time (or the time that it takes V_{OUT} to stabilize at a new frequency) increases with increased values of C1. In turn, the value of C1 is affected by the value of R1, in the following relationship:

$$C1 = V3MAX / R1 \times VCC \times f_{FULL SCALE}$$

where $f_{FULL SCALE}$ is the full-scale input frequency. Because of this relationship between R1 and C1, there is a compromise or tradeoff among ripple, response time, and linearity.

When the maximum acceptable ripple is determined, the value of C2 can be found using the following equation:

$$C2 = V_{CC} / 2 \times C1/V_{RIPPLE} \; 1 - V_3 / R_1 I_2$$

Note that the type of capacitor used for C1 determines the accuracy of the FVC circuit over the temperature range. (Figure 5–3 shows tachometer output as a function of temperature using a similar circuit.) Also note that the LM2907, operating from a fixed external supply, has a negative temperature coefficient (TC) that enables the IC to be used with capacitors having a positive TC, and thus provides overall stability. In the case of the LM2917, the internal zener supply voltage has a positive TC. This causes the overall tachometer output to have a very low TC, and requires that the capacitor TC be balanced by the TC of R1.

5.1.4 Internal Zener Regulation

When the output voltage must be maintained constant in the presence of unregulated supply-voltage variations, use the LM2917 with built-in zener reference and an external dropping resistor. (Note that the internal zener has an 11-ohm source resistance.) Use the graph of Fig. 5–4 to find the correct value for the dropping resistor.

In choosing the dropping resistor (connected from the unregulated supply to the IC at pin 8), note that the internal circuitry requires about 3 mA at the voltage level (7.6 V) supplied by the zener. At low supply voltages, there must be some current flowing in the resistor (above the 3-mA circuit current) to operate the zener. As an example, if the raw supply varies from 9 V to 16 V, a resistance of 470 ohms will minimize zener variations to 160 mV. If the resistance is below 400 ohms, or above 600 ohms, the zener variation rises above 200 mV for the same unregulated supply variation. Also, as always, make certain that the power dissipation of the IC is not exceeded at the highest supply voltage.

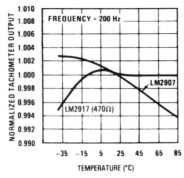

Figure 5–3.
Tachometer output as a
function of temperature
(National Semiconductor,
*Linear Applications
Handbook*, 1994, p. 354)

TL/H/7451–32

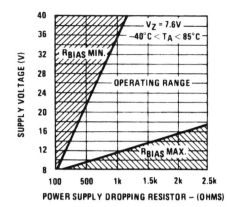

Figure 5–4.
Zener-regulator bias-resistor range (National Semiconductor, *Linear Applications Handbook,* 1994, p. 349)

POWER SUPPLY DROPPING RESISTOR – (OHMS)

TL/H/7451–7

5.1.5 Input Circuit Considerations

Figure 5–5 shows some typical input-circuit configuration. The following guidelines should be considered when using any of the input circuits.

Figure 5–5(a) shows how the ground-referenced input of the LM2907 provides for direct coupling to transformer inputs, or variable-reluctance pickups.

Figure 5–5(b) shows an AC-coupled input where the frequency-signal source or output does not go below ground. This approach is suitable for use with phototransistors and optical pickups.

Figure 5–5(c) shows a bandpass-filtered input where the signal source is noisy. An example is a tachometer operating from breakerpoints on an older ignition system. Keep in mind that the minimum input signal required by the LM2907 is 30 mVp-p, but this signal must be able to swing at least 15 mV on either side of the inverting input.

The maximum signal which can be applied to the LM2907 input is ±28 V. The input-bias current is typically 100 nA. A path to ground must be provided for the current, either through the source or by other means as shown.

Figure 5–5(d) shows an above-ground sensing circuit for use with the 14-pin versions of the LM2907 and LM2917. Here, the inverting input is biased by an external source. This permits the circuit to operate with input signals that do not go to ground, but are referenced at higher voltages. The method also increases the noise immunity where large signal levels are available, but large noise signals are also present.

Figure 5–5(e) shows a balanced-bias input with high common-mode rejection. This circuit take full advantage of the differential-input stage. With the balanced-bias input, the effective common-mode rejection is virtually infinite (because of the usual high common-mode rejection of a differential input stage, and the effect of input hysteresis).

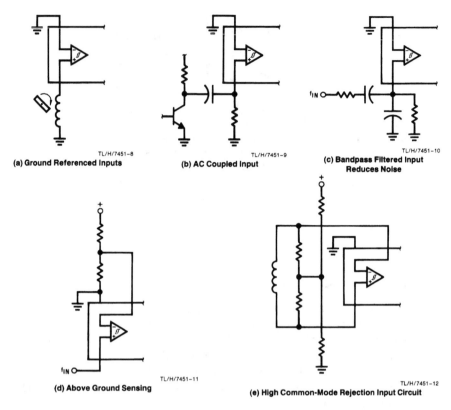

TL/H/7451-8
(a) Ground Referenced Inputs

TL/H/7451-9
(b) AC Coupled Input

TL/H/7451-10
**(c) Bandpass Filtered Input
Reduces Noise**

TL/H/7451-11
(d) Above Ground Sensing

TL/H/7451-12
(e) High Common-Mode Rejection Input Circuit

Figure 5–5. Typical input-circuit configurations (National Semiconductor, *Linear Applications Handbook,* 1994, p. 349)

5.1.6 Output Circuit Considerations

Figure 5–6 shows some typical output-circuit configurations. The following guidelines should be considered when using any of the output circuits.

The output circuits of the LM2907/LM2917 can be interfaced with a wide variety of loads. This flexibility results from the availability of both the collector and emitter of the output transistor (which is capable of driving up to 50 mA of load current). When the non-inverting input is higher than the inverting input, the output transistor is turned on, and can be used to drive loads to either the positive or the negative supply with the emitter or collector, respectively, connected to the other supply.

For example, as shown in Fig. 5–6(a), a simple speed switch can be constructed so that the speed signal taken from the FVC is compared to a reference voltage derived from a resistive divider. When the speed signal exceeds the reference, the output transistor turns on the LED in the load. The small current-limiting resistor in series with the LED protects the output circuit.

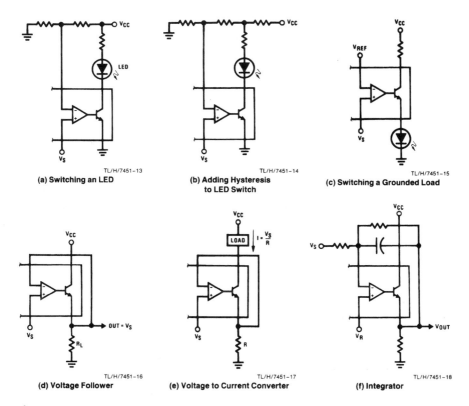

Figure 5-6. Typical output-circuit configurations (National Semiconductor, *Linear Applications Handbook,* 1994, p. 350)

Figure 5–6(b) shows how hysteresis can be added to the speed switch. The circuit of Fig. 5–6(a) has no hysteresis (the turn-on and turn-off speed voltages are essentially equal). In cases where speed might fluctuate at a high rate, and a flashing LED is objectionable, use the circuit of Fig. 5–6(b) where the switch-on speed is above the switch-off speed by a controlled amount.

Figure 5–6(c) shows how a grounded load can be switched. In this case, the current-limiting resistor is in the collector of the output transistor. The base current of the output transistor is limited by an internal 5-k base resistor. This raises the output resistance so that the output swing is reduced as full load.

Figure 5–6(d) shows the output connected as a voltage follower. (The op-amp/comparator stage is internally compensated for unity-gain feedback configuration.) By directly connecting the emitter output to the non-inverting input, the op amp becomes a voltage follower. Of course, the op amp can also be operated as an amplifier, integrator, active filter, or in any other normal op-amp configuration.

Figure 5–6(e) shows a unique configuration that is not available with standard op amps. Here, the collector of the output transistor is used to drive a load, where the current is proportional to the input voltage. In effect, the circuit is operating as a voltage-to-current converter. This is ideal for driving remote signal sensors and moving-coil galvanometers.

Figure 5–6(f) shows how an active integrator can be used to provide an output that falls with increasing speed.

5.1.7 Transient Protection Circuits

Figure 5–7 shows some transient protection schemes for the LM2907/LM2917. Such protection is required because many applications use unregulated power supplies that tend to expose the ICs to potentially damaging transients on the power-supply line. The following guidelines should be considered when using any of the transient protection circuits.

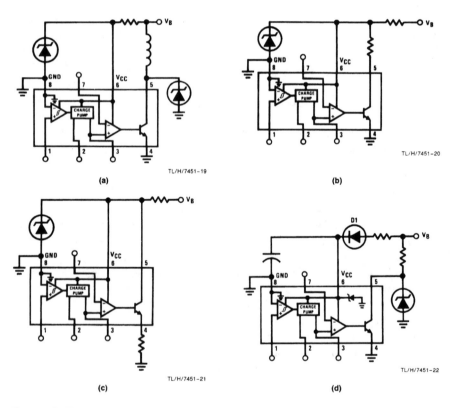

Figure 5–7. Transient protection circuits (National Semiconductor, *Linear Applications Handbook*, 1994, p. 351)

Transients pose two special problems in automotive applications. First, there is the load-dump transient. This occurs when a dead battery is being charged at a high current and the battery cable is disconnected. Under this condition, the current in the alternator inductance produces a positive transient on the line of 60 V to 120 V. The second transient is called field decay. This occurs when the ignition is turned off, and the energy stored in the field winding of the alternator causes a 75-V transient on the ignition line.

Figure 5–7(a) shows a situation where the power supply to the LM2907 is provided through a dropping resistor and regulated by an external zener diode Z1, but the output drive operates from the full supply voltage. In this case, a separate protection zener must be provided if the voltage on the power line is expected to exceed the maximum rated voltage of the LM2907.

In the circuits of Figs. 5–7(b) and 5–7(c), the output transistor is required only to drive a simple resistive load. No secondary-protection circuits are required. (Note that the dropping resistor to the zener must supply current to the output circuit.) With either of these circuits, reverse supply protection is provided by the forward-biased zener diode. The zener should be a low forward-resistance unit to limit the maximum reverse voltage applied to the IC. Excessive reverse voltage on the IC can cause high currents to be conducted by the substrate diodes (and can cause permanent damage to the IC). Up to 1 V negative can generally be tolerated with either version of the IC. With the LM2917, which has an internal zener, the circuit is self-protection (depending on the size of the dropping resistor used).

Figure 5–7(d) shows a transient-protection circuit where large negative transients are anticipated. The blocking diode D1 is connected between a dropping resistor and the V_{CC} pin. During negative transients, D1 is reverse biased and prevents reverse currents from flowing into the IC. If these transients are short and the value of the capacitor at pin 2 is high enough, power to the IC can be sustained without damage. This circuit is useful in preventing change-of-state or change-of-charge in speed-sensing systems.

5.1.8 FVC Overspeed Switch

Figure 5–8 shows the FVC connected as an overspeed sensor or switch. With this circuit, the load is energized when the input frequency exceeds a given value. The equation involved is:

$$f_{IN} > 1/(2C1R1)$$

where f_{IN} is the input frequency and C1/R1 are the components at pins 2 and 3.

Using the values shown (C1 = 0.3 mF, R1 = 100 k), the load is energized when the input frequency exceeds 16.6 Hz.

A typical use for this circuit is one in which a warning must be given when a certain maximum speed is exceeded. For example, the load can be an alarm buzzer or lamp that goes on when a gear wheel (such as that shown in Fig. 5–2) rotates past the pickup faster than a desired maximum.

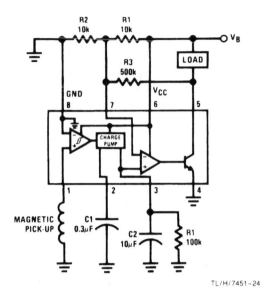

Figure 5-8.
FVC connected as an
overspeed sensor or
switch (National Semi-
conductor, *Linear Appli-
cations Handbook,* 1994,
p. 352)

TL/H/7451-24

Because the reference level on the comparator (pin 7) is set by two equal resistors tied to VCC, the reference is a function of supply voltage, as is the output from
the charge pump. As a result, there is no need for elaborate supply regulation. Also,
the frequency at which switching occurs is independent of supply voltage.

5.1.9 FVC Overspeed Latch

Figure 5-9 shows the FVC connected as an overspeed latch or shutdown circuit. With this configuration, the FVC output increases with frequency until a certain
set point is reached. The output then goes up to VCC (or some voltage near VCC, depending on the drop across the output transistor), and remains at VCC until the power
is turned off. This latch function is done by connecting the emitter of the output transistor back to the non-inverting input of the comparator (pin 3 to pin 4). As shown by
the equations, the set-point frequency is determined by external components.

5.1.10 FVC with Analog Display

Figures 5-10 and 5-11 show FVCs connected to provide an analog display or
readout of input frequency. Both circuits use moving-coil meters calibrated directly
in frequency (or with some other convenient scale). The equations for the selection of
external components are the same as described for the basic FVC circuit. The circuit
of Fig. 5-10 can be calibrated by adjustment of R2.

In both circuits, the output transistor provides current drive to the moving-coil
meter. This avoids temperature tracking problems with varying meter resistance, and

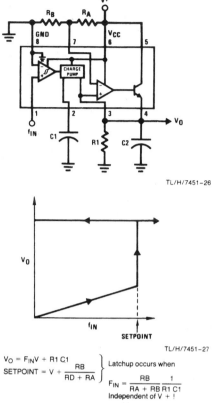

TL/H/7451-26

Figure 5-9.

FVC connected as an overspeed latch (National Semiconductor, *Linear Applications Handbook,* 1994, p. 353)

TL/H/7451-27

$V_O = F_{IN}V + R1 C1$

$\text{SETPOINT} = V + \dfrac{RB}{RD + RA}$ } Latchup occurs when

$F_{IN} = \dfrac{RB}{RA + RB} \dfrac{1}{R1 C1}$
Independent of V + !

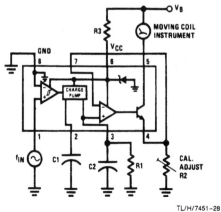

Figure 5-10.

FVC with analog display of frequency (National Semiconductor, *Linear Applications Handbook,* 1994, p. 353)

TL/H/7451-28

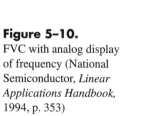

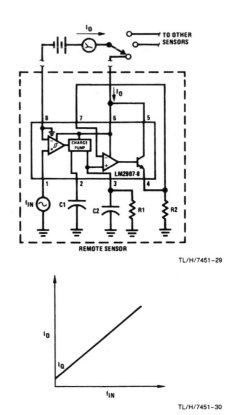

TL/H/7451–29

Figure 5–11.
FVC with remote analog
display of frequency (National Semiconductor, *Linear Applications Handbook,* 1994, p. 354)

TL/H/7451–30

permits high-resistance instruments to be driven accurately with relatively large voltages.

The LM2917 version (with internal zener) is used in Fig. 5–10 to provide a regulated current to the display instrument. The onboard 7.6-V zener is compatible with car and boat batteries, and permits the moving-coil instrument to use the full battery voltage (12 V) for deflection. This enables high-torque meters to be used, and is particularly useful in high-vibration environments such as boats and motorcycles.

The circuit of Fig. 5–11 is used where the sensor and display device are not at the same location. This circuit cuts down the number of wires needed between the sensor and display. The output current is conducted along the supply line so that a local current-sensing device in the supply can be used to get a direct reading of the frequency at the remote location (where the electronics might also be situated). The small zero-speed offset, because of the device quiescent current, can be compensated by offsetting the zero on the display device. This circuit also permits one display device to be shared among several inputs.

5.1.11 FVC Tachometer

Figure 5–12 shows the FVC connected as a tachometer for a gasoline engine. In this case, the input is driven from the spark coil. The tachometer can be set up for any number of cylinders by linking the appropriate timing resistor as shown. The circuit is calibrated by the variable resistor at pins 5 and 10. The zener and series resistor provide protection against transients, which are common in automotive electrical systems.

5.1.12 FVC Motor-Speed Controls

Figures 5–13, 5–14, and 5–15 show FVCs connected to provide speed control for small DC motors. Such motors are available with a built-in AC tachometer (such as from TRW Globe Motors, 2275 Stanley Avenue, Dayton, Ohio, 45404). The combination of a motor with built-in tachometer and an FVC IC provides a low-cost speed control system.

In the circuit of Fig. 5–13, the tachometer drives the non-inverting input of the comparator up toward the preset reference level. When that level is reached, the out-

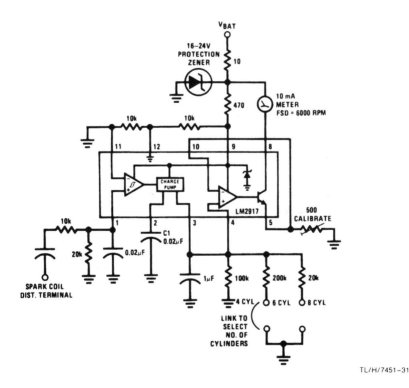

TL/H/7451–31

Figure 5–12. FVC connected as a tachometer for gasoline engine (National Semiconductor, *Linear Applications Handbook,* 1994, p. 354)

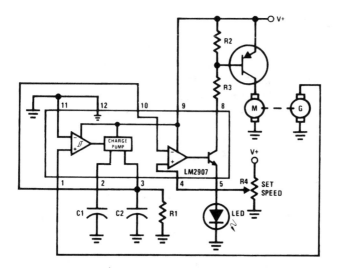

TL/H/7451–33

Figure 5–13. FVC as a motor-speed control (National Semiconductor, *Linear Applications Handbook,* 1994, p. 355)

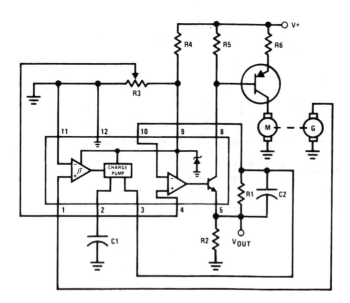

TL/H/7451–34

Figure 5–14. FVC as a proportional motor-speed control (National Semiconductor, *Linear Applications Handbook,* 1994, p. 355)

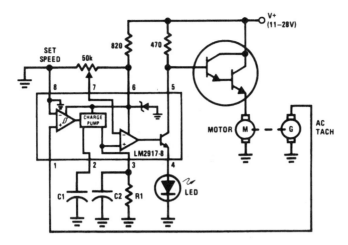

TL/H/7451-35

Figure 5-15. FVC as a shunt motor-speed control (National Semiconductor, *Linear Applications Handbook,* 1994, p. 356

put is turned off and the power is removed from the motor. As the motor slows down, the voltage from the charge-pump output falls and power is restored. Speed is maintained by operating the motor in a switching mode. Hysteresis can be provided to control the rate of switching.

The circuit of Fig. 5-14 provides proportional control of the motor. In this circuit, the charge-pump integrator is connected to provide feedback around the op amp. The output voltage for zero speed is equal to the reference voltage set up on the pot in the non-inverting input. As speed increases, the charge pump puts a charge into capacitor C2, and causes the output V_{OUT} to fall in proportion to speed.

The current of the output transistor is used to provide an analog drive to the motor. As the motor speed approaches the reference level, the current is reduced (in proportion) so that the motor gradually comes up to speed, and is maintained without operating the motor in a switching mode. This configuration is particularly useful in situations where the electrical noise generated by the switch-mode operation (Fig. 5-13) is objectionable. However, the load regulation of the Fig. 5-14 circuit is poor.

The circuit of Fig. 5-15 uses the FVC as a shunt-mode regulator. The LED at pin 4 indicates when the FVC is in regulation. Note that the equations for the selection of C1, C2, and R1 are the same as described for the basic FVC circuit.

5.1.13 FVC Staircase Generator

Figure 5-16 shows the FVC connected as a staircase generator. Such a circuit can be used to indicate position in a series of counts. In this circuit, the timing resistor is removed from pin 3 so that the output current produces a staircase instead of a fixed DC level. If the magnetic pickup (or other frequency input) senses passing

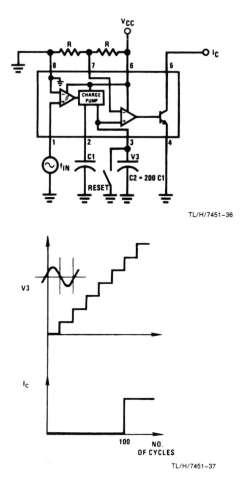

Figure 5-16.
FVC as a staircase generator (National Semiconductor, *Linear Applications Handbook,* 1994, p. 356)

TL/H/7451-36

TL/H/7451-37

notches or items, the staircase signal is generated, and can be compared with a reference to initiate a switching action when a specified count is reached. The circuit of Fig. 5–16 counts up 100 input pulses and then switches on the output transistor. Examples of this application can be found in automated packaging operation or in line printers.

5.1.14 FVC with Capacitive Input

Figure 5–17 shows the FVC connected for operation with two inputs. A capacitive input (a variable capacitor driven by a mechanical positioning device) is applied at pin 2 in place of the usual fixed capacitor C1. A fixed input (in this case a fixed 60-Hz signal taken from the power line) is applied to the frequency input at pin 1. The result is an output that is proportional to the capacitance value. In turn, the capaci-

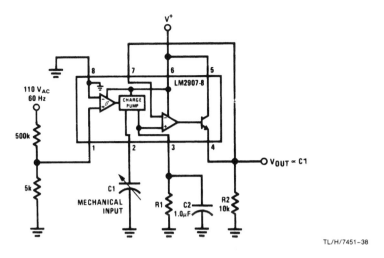

Figure 5–17. FVC connected for operation with capacitive inputs (National Semiconductor, *Linear Applications Handbook,* 1994, p. 357)

tance value represents mechanical position (rotation and such). If the input at pin 1 is made variable, the output is then proportional to the product of the two variables (at pins 1 and 2). Such an arrangement makes it possible to multiply two variables.

The circuit of Fig. 5–17 could be used in flow-measurement systems. For example, the input frequency (at pin 1) can represent a variable that depends on flow rate, such as a signal generated from a paddle wheel, propeller, or vortex sensor. The capacitor input (at pin 2) can provide an indication of orifice size or aperture size, such as a throttle body. The product of these two factors indicates volume flow. A thermistor can be added to R1, thus converting volume flow to mass flow. As can be seen, a combination of many inputs (including control voltage on the supply) can be used to provide complex multiplication of analog values, with independent control of the variables.

5.1.15 Using FVCs with PLLs and VCOs

Figure 5–18 shows how FVCs can be used to increase the capture range of PLLs (phase-locked loops). Although there are many PLLs available in IC form, they often have a narrow capture range and hold-in range. (PLLs are discussed further in Section 5.2.)

In the circuit of Fig. 5–18, the FVC (an LM2907) initially puts the VCO (voltage controlled oscillator) at approximately the right frequency to match the input frequency. The phase detector is then used to close the gap between the VCO and the input frequency by exerting control on the summing point. Using this system, a wide-range phase loop can be developed, provided that there is proper tracking between the FVC and the VCO.

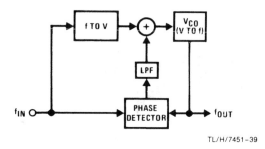

Figure 5-18.
FVC used to increase
capture range of PLLs
(National Semiconductor,
*Linear Applications
Handbook,* 1994, p. 357)

TL/H/7451-39

Figure 5–19 shows how FVCs can be used to improve the linearity of VCOs. In this case, the LM2709 is used as a feedback-control element. The output frequency of the VCO is converted back to a voltage that is compared with the input voltage. This configuration can often be a lower cost solution to providing a linear VCO than by working directly on the VCO in the open-loop mode.

5.1.16 FVCs as Zero-Crossing Detectors

Figures 5–20 and 5–21 show how FVCs can be used as zero-crossing detectors. Such circuits can provide inputs to microprocessors and other digital control systems to indicate position or time from some mechanical source.

At each zero-crossing of the input signal to the circuit of Fig. 5–20, the charge pump changes the state of capacitor C1, and provides a one-shot pulse into the zener diode at pin 3. The width of pulse is controlled by the internal current of pin 2 and the size of C1, as well as by the supply voltage. Because a pulse is generated by each zero crossing of the input signal, the circuit is a "two-shot" configuration, and doubles the frequency presented to the microprocessor or other control element.

If frequency doubling is not required, and a square-wave output is preferred, use the "one-shot" configuration shown in Fig. 5–21. In this circuit, the output swing is the same as the swing on pin 2 (equal to one-half the supply voltage, starting at one base-emitter voltage below one-quarter of the supply, and going to one base-emitter

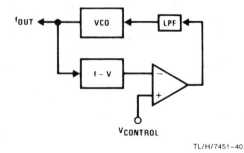

Figure 5-19.
FVC used to improve
linearity of VCOs (National Semiconductor,
*Linear Applications
Handbook,* 1994, p. 357)

TL/H/7451-40

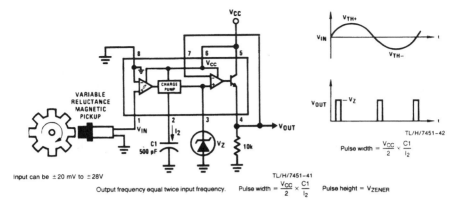

Figure 5–20. FVC as a two-shot zero-crossing detector (National Semiconductor, *Linear Applications Handbook,* 1994, p. 358)

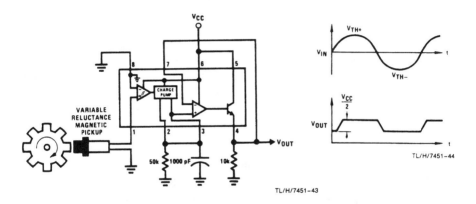

Figure 5–21. FVC as a one-shot zero-crossing detector (National Semiconductor, *Linear Applications Handbook,* 1994, p. 358)

voltage below three-quarters of the supply). The output can be increased up to the full output-swing capability by removing the negative feedback around the op amp.

5.1.17 FVC as an Analog-to-Digital Converter

Figure 5–22 shows how an FVC can be used as an analog-to-digital converter (ADC) under control of a digital processor (typically a microprocessor). Note that this circuit is similar to the staircase generator of Fig. 5–16, except that the Fig. 5–22 circuit is under digital control.

To start a conversion cycle, the processor generates a reset pulse to discharge the integrating capacitor C2. Each complete clock cycle generates a charge and discharge cycle on C1. This results in two steps per cycle being added to C2. As the volt-

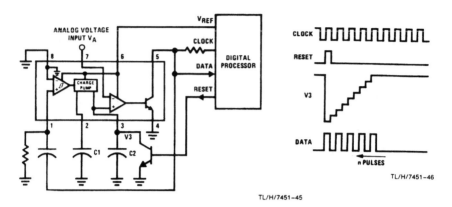

Figure 5–22. FVC as an ADC (National Semiconductor, *Linear Applications Handbook*, 1994, p. 358)

age on C2 increases, clock pulses are returned to the processor. When the voltage on C2 steps above the analog voltage, the data line is clamped and C2 stops charging. By counting the number of clock pulses received after the reset pulse, the processor is loaded with a digital measure of the input voltage. By making C2/C1 = 1,024, an 8-bit ADC is formed.

5.1.18 FVCs in Anti-Lock Braking Systems

Figs. 5–23, 5–24, and 5–25 show how FVCs can be used in anti-lock braking systems (or anti-skid systems). In automotive applications, such systems often use variable-reluctance pickup sensors on the wheels to provide inputs to the control circuits. The circuits shown here permit the system designer to use the average signal from each of the two wheels on a given axle, or the lower of two speeds, or the higher of two speeds.

In the select-low circuit of Fig. 5–23, the input frequency from each wheel sensor is converted to a voltage in the normal manner. The op-amp/comparator circuit is connected with negative feedback (using a diode in the loop) so that the amplifier can only pull down on the load and not pull up. In this way, the outputs from the two devices can be joined together, and the resultant output voltage represents the lower of the two input speeds.

In the select-high circuit of Fig. 5–24, the output emitter of the op amp provides the pull-up required to produce a configuration where the output voltage represents the higher of the two speeds. The select-average circuit of Fig. 5–25 saves components by allowing the two charge pumps to operate from a single RC network. One of the amplifiers is needed to buffer the output, and to provide a low impedance output that represents the average of the two input frequencies. The second amplifier is available for other applications.

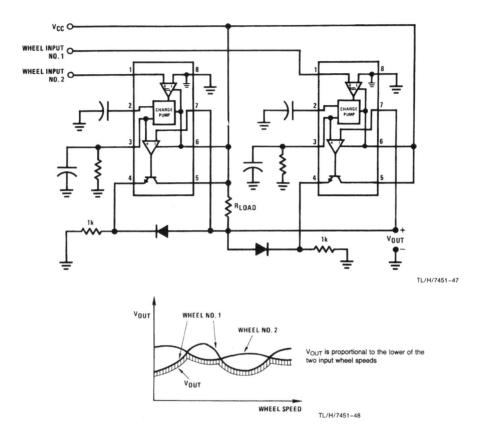

TL/H/7451–47

TL/H/7451–48

Figure 5–23. FVC used in a select-low anti-lock braking system (National Semiconductor, *Linear Applications Handbook,* 1994, p. 359)

5.1.19 FVCs in Transmission and Clutch-Control Systems

Figure 5–26 shows how FVCs can be used in transmission and clutch-control systems. This circuit operates electric clutches on the transmission to eliminate slip during cruise (and thus provide better fuel economy).

Magnetic pickups are connected to input and output shafts of the transmission, respectively, and provide frequency inputs f1 and f2 to the circuit. Frequency f2 (representing the output-shaft speed) is also a measure of vehicle road speed. As a result, the number-2 FVC provides a voltage that is proportional to road speed. This voltage is buffered by the op amp in the number-1 FVC to provide a speed output (V_{OUT1}) at pin 4.

The input shaft provides charge pulses (at the rate of 2f1) into the inverting input of the number-2 FVC. This input has the integrating network R1-C3 connected to

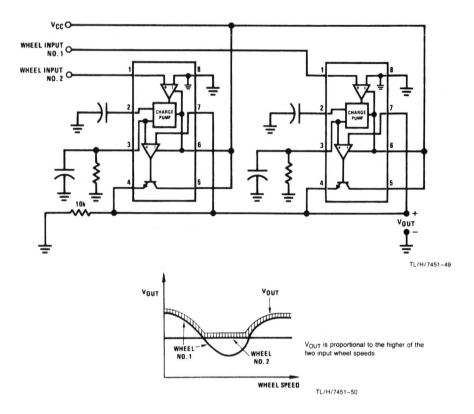

TL/H/7451–49

TL/H/7451–50

Figure 5–24. FVC used in a select-high anti-lock braking system (National Semiconductor, *Linear Applications Handbook,* 1994, p. 360)

the output so that the change pulses are integrated, and provide an inverted output voltage that is proportional to the input speed. As a result, the V_{OUT2} is proportional to the difference between the two input frequencies.

With these two signals (the road speed and the difference between road speed and input shift speed) it is possible to develop an number of control functions, including the electronic clutch and a complete electronic transmission control.

Note that in the configuration shown, it is not possible for V_{OUT2} to go below zero. As a result, there is a limitation to the output swing in this direction. This problem can be overcome by returning R3 to a negative bias supply instead of to ground.

5.2 Using PLLs as FVCs

Because PLLs can operate over a wide frequency range (typically two or three decades), and can provide a voltage output that responds quickly to frequency

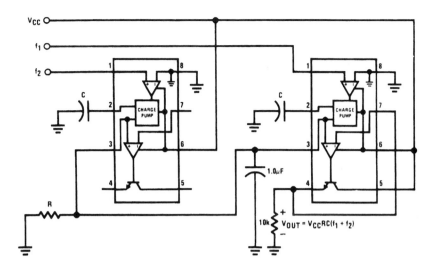

Figure 5–25. FVC used in a select-average anti-lock braking system (National Semiconductor, *Linear Applications Handbook,* 1994, p. 360)

changes, they can be used as FVCs. When PLLs are used as FVCs, the compromise between large ripple versus slow response is eliminated. Also, the linearity of such an FVC will be as good as the linearity of the VFC used in the circuit (easily better than 0.01%). However, there is a disadvantage in that an FVC formed by a PLL requires a clean, noise-free input frequency, such as a square wave or pulse train.

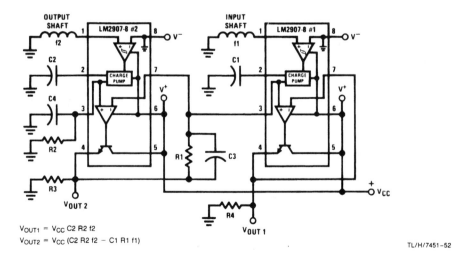

Figure 5–26. FVC used in transmission and clutch-control systems (National Semiconductor, *Linear Applications Handbook,* 1994, p. 361)

5.2.1 Basic PLL-Type FVC

Figure 5–27 shows the basic circuit of a PLL used as an FVC. Note that the frequency/phase detector uses D-type FFs instead of the usual quadrature detector. The following is a description of the major circuit functions in Fig. 5–27.

5.2.2 Frequency and Phase Detector

When the frequency input (F_{IN}) to the circuit of Fig. 5–27 is larger than F_2, Q1 is forced high most of the time, and provides a positive error signal (through CR3, CR4, CR5, and CR6) to the error integrator. If the input frequency and F_2 are the same, but the rising edges of F_{IN} lead the rising edges of F_2, the duty cycle of Q1 = HI will be proportional to the phase error. As a result, the error signal fed to the integrator decreases to nearly zero, when the loop has achieved phase-lock, and the phase error between F_{IN} and F_2 is zero.

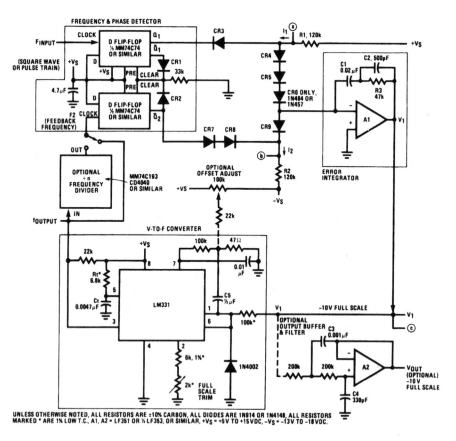

Figure 5–27. Basic PLL used as an FVC (National Semiconductor, *Linear Applications Handbook*, 1994, p. 399)

Actually, when the phase error is zero, Q_1 produces 30-ns positive pulses, at the same time that Q_2 puts out 30-ns negative pulses, and the net effect (as seen by the integrator) is zero net change. The 30-ns pulses at Q_1 and Q_2 enable both FFs to be cleared, and prepared for the next cycle.

The phase-detector action is substantially the same as that of an MC4044 phase detector, but the MM74C74 is cheaper and uses less power. The MM74C74 is fast enough for frequencies below 1 MHz. At higher frequencies, a DM74S74 can be used in the same way, with very low delays.

5.2.3 Error Integrator

The error integrator takes in the current from R1 or R2, as gated by the Q_1 and $\overline{Q_2}$ outputs of the FF. For example, when F_{IN} is higher, and Q_1 is H_I, I_1 flows through CR4, CR5, and CR6, and causes the integrator output to go more negative. This is the direction to make the VFC run faster, and bring F_2 up to F_{IN}.

Note that A1 does not merely integrate the current in C1. The resistor R3 in series with C1 causes a phase lead in the loop response, which is essential to loop stability. The small capacitor C2 across R3 is not essential, but offers improved settling at the voltage output.

5.2.4 VFC Circuit

The V_1 output of the integrator is fed to the VFC. An LM331 is used in this example. The LM331 runs on a single supply, and responds quickly with a linearity of better than 0.05%, even though an op amp is not used. The output of the VFC is fed back to the F_2 input of the frequency and phase detector, either directly or through an (optional) frequency divider. Any number of standard frequency dividers, such as an MM74C193, CD4029, or CD4018, can be used (within reasonable limits). A divider of 2, 3, 10, or 16 is often used.

5.2.5 Unfiltered Output

The V_1 output is proportional to the input frequency (limited only by the linearity of the VFC). As a result, the V_1 output can be used directly as the FVC output. However, during the brief period when the FF is clearing itself, there will be small noise glitches at the output of A1. The RMS value of this noise is small (typically 0.5 to 5 mV), but the peak amplitude could be 10 to 100 mV, and can be a problem in some systems. The noise problem can be corrected by filtering.

5.2.6 Filtered Output

No additional filtering can be added to the main-loop path. Any further delay in the signal to the VFC could cause loop instability. However, the FVC output can be taken from a separate filter and buffer that operates as a branch path. A2 provides a simple 2-pole active filter that cuts the steady-state ripple and noise down below 1 mV peak-to-peak. Also, the A2 output settles faster than the A1 output (unfiltered V1). Typically, A2 settles in about 2 ms, whereas A1 settles in about 12 to 14 ms.

5.2.7 Proportional Current Source

Figure 5–28 shows a proportional-current source for the basic circuit of Fig. 5–27. Such a current source might be required when operating at lower frequencies. The basic circuit operates properly over a frequency range of about 3:1. However, if the frequency is decreased below 3 kHz, the loop gain becomes excessive, and currents I_1 and I_2 are large enough to cause loop instability.

The loop gain increases at lower frequencies because a given initial phase error causes the fixed current from R1 or R2 to be integrated for a longer time. In turn, this causes a larger change at the integrator output, and a larger change of frequency. When the frequency is thus corrected, and the period of one cycle is changed, the circuit can be overcorrected. Also, the phase error on the next cycle might be as large as (or larger than) the initial phase error, but with the sign reversed.

To avoid overcorrection, and to maintain loop stability at lower frequencies (0.5 to 1 kHz), raise the value of R1 and R2 to 1.5 M, rather than the 120 k shown. However, the increase in R1 and R2 slows the response to a step change. If the slower response cannot be tolerated, use the circuit of Fig. 5–28 to produce currents that are proportional to F_{IN}.

The circuit of Fig. 5–28 is used in place of R1/R2, and provides good loop stability over a 30:1 frequency range, from 330 Hz to 10 kHz. For best results at this range, change the value of damping resistor R3 (Fig. 5–27) from 47 k to 100 k. To cover frequency ranges wider than 30:1, use one of the other circuits described in this section. Keep in mind that if the frequency range need only be in the 2:1 to 3:1 range, use the basic circuit of Fig. 5–27 without change.

5.2.8 Frequency Multiplication

In some systems, a frequency multiplier is needed to provide an output frequency many times (n times) higher than the input. By inserting a divide-by-n

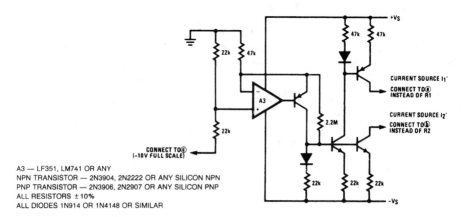

Figure 5–28. Proportional-current source for basic PLL (National Semiconductor, *Linear Applications Handbook*, 1994, p. 402)

frequency divider in the feedback loop, the basic circuit can be modified to accommodate the frequency change. It is also possible to insert a divide-by-m frequency divider ahead of the frequency input (to provide correct scaling), and the output frequency will then be F_{IN} (n/m).

To get good loop stability in a frequency multiplier where n = 2, remember that a 20-kHz VFC followed by a divide-by-2 circuit has exactly the same loop response and stability needs as a 10 kHz VFC. This is because the resultant circuit is a 10-kHz VFC, even though the circuit provides a useful 20-kHz output. As a result, the frequency of the F2 (minimum and maximum) determines what loop gains and loop-damping components are needed.

To accommodate a 1-kHz VFC loop, simply make C1 and C2 10 times larger than the values of Fig. 5–27. If C3, C4, and C5 are used, increase these capacitor values by a factor of 10 also. To accommodate a 100-Hz VFC loop, increase the values by a factor of 100.

5.2.9 Single-Supply PLL/FVC

Figure 5–29 shows a single-supply PLL used as an FVC. Because of the single supply, this circuit is well suited for battery operation. The circuit functions accurately over a 10:1 frequency range from 1 kHz to 10 kHz, but will not respond as quickly as the basic circuit in Fig. 5-27. This slower speed is caused by the use of a CD4046 frequency detector. Here's why.

When the leading edge of an F_{IN} signal occurs ahead of a feedback pulse, pin 13 of the CD4046 pulls up on C1 through R1. This current cannot be controlled or manipulated over as wide a range as I1 in Fig. 5–27. As a result, the response of the PLL is never as smooth nor fast-settling as the basic PLL. However, the circuit of Fig. 5–29 is still generally superior to most FVCs.

The detector feeds a current to the integrator (R1/C1) and buffer A1. The optional A3 can provide a filtered output if required. A2 controls Q1, drawing a current from C6. This current is proportional to V2. The LM331 acts as a current-to-frequency converter, and produces an output (f_{OUTPUT}) that is proportional to the collector current of Q1.

As in the case of the basic circuit, the PLL of Fig. 5–29 can be used as a quick and/or quiet FVC, or as a frequency multiplier. One of the most important uses of an FVC is to demodulate the frequency of a VFC, which can be situated at a high common-mode voltage (isolated by photoisolators), or to recover a telemetered signal. An FVC of this sort can provide good bandwidth for demodulating such a signal.

5.2.10 Precision PLL

Figure 5–30 shows a precision PLL used as an FVC. This circuit is similar to the basic circuit of Fig. 5–27, but with the following improvements.

- The flip-flops in the detector have a gate G1 to provide quicker clearing after each pulse.

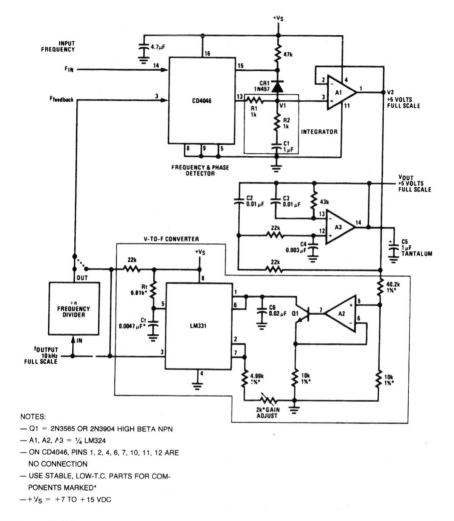

Figure 5–29. Single-supply PLL used as an FVC (National Semiconductor, *Linear Applications Handbook,* 1994, p. 403)

- The currents that A1 integrate are steered through Q1, Q2, Q3, and Q4. These transistors are quicker than diodes, yet have much lower leakage.
- The VFC uses A2 as an op-amp integrator to get better than 0.01% non-linearity (maximum).
- Gate G2 is recommended as an inverter, to invert the signal on pin 3 of the LM331 VFC. This avoids delay and improves loop stability (theoretically).
- Amplifier A4 is included as an (optional) limiter to prevent voltage V1 from going positive. This will facilitate quick startup and recovery from overdrive conditions.

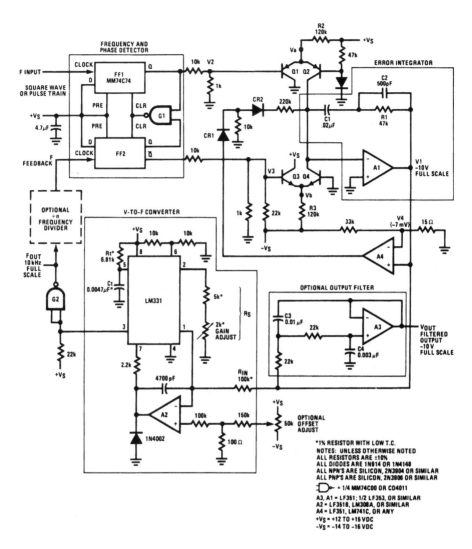

Figure 5–30. Precision PLL used as an FVC (National Semiconductor, *Linear Applications Handbook,* 1994, p. 404)

Figure 5–31 shows a wide-range current pump for the circuit of Fig. 5–30. This pump provides an output current that is proportional to –V1 (within 10 to 15%) over a 3-decade range. The 22-M resistors prevent the current from shutting off in case –V1 becomes positive. (This is probably unnecessary if A4 is used.)

For best results over a full 3-decade range (11 kHz to 9 Hz), use A4, delete the four 22-M resistors, and insert the parallel diode and 470-k resistor in series with resistor R_G, as shown. This will give good stability at all frequencies. (However, stabil-

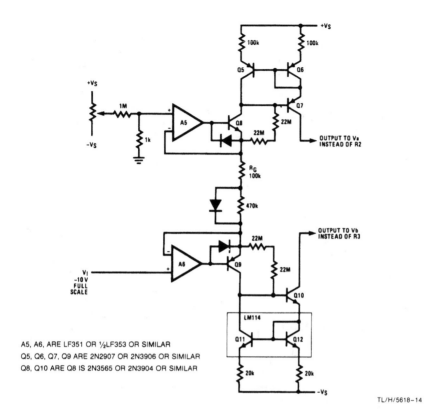

Figure 5–31. Wide-range current pump for the precision PLL (National Semiconductor, *Linear Applications Handbook,* 1994, p. 405)

ity cannot be extended below 1/1500 of full scale, without extensive circuit modifications.)

The circuit of Fig. 5–30 has been used to test VFCs. For example, the circuit can force the LM331 to run at a crystal-controlled frequency (established as the F input), while the output is measured with a 6-digit (1 ppm nonlinearity, maximum) digital voltmeter.

Such a test scheme has obvious advantages over the usual method of applying a voltage and trying to read a frequency, especially at low frequencies. For example, measuring a 50-Hz signal with ±0.01 Hz resolution cannot be done (even with the most powerful computing counter-timer) as accurately, quickly, and conveniently as measuring the voltage output from a PLL.

5.2.11 VFC Function Generator

Figure 5–32 shows a discrete-component VFC that produces sine-wave, square-wave, and triangular outputs, where frequency is determined by the input

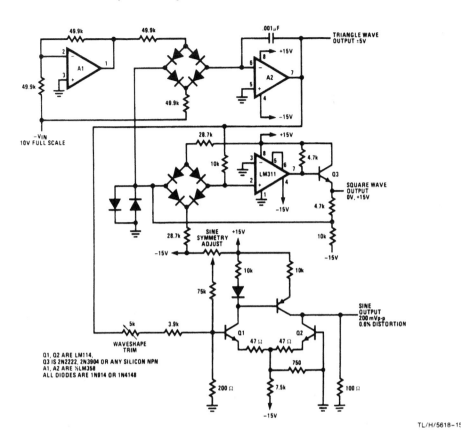

Figure 5–32. VFC function generator (National Semiconductor, *Linear Applications Handbook,* 1994, p. 406)

voltage (a V_{IN} of –10 V full-scale, in this case). Using the configuration shown, the linearity and frequency stability are about 0.2%. If this is not satisfactory, both linearity and stability can be increased by combining the circuit of Fig. 5–32 with one of the PLLs described in this section (Figs. 5–27, 5–29, or 5–30).

To make a practical circuit of the combination, simply connect the sine VFC of Fig. 5–32 into the PLL, instead of the LM331 VFC. Then use a precise, low-drift VFC (based on the LM331) to establish the F_{IN} to the PLL. If the sine VFC output drifts, the PLL integrator will compensate, and provide a stable, linear output. Refer to Section 5.6 for further information on using the LM331.

5.3 FVC with Sample-and-Hold

Figure 5–33 shows a basic FVC using the LM331. Figure 5–34 shows an improved version of the circuit using an LM331 and an LF398 sample-and-hold (S/H)

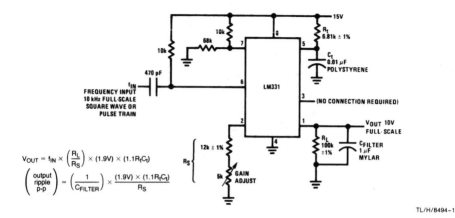

TL/H/8494-1

Figure 5–33.   Basic FVC using an LM331 (National Semiconductor, *Linear Applications Handbook*, 1994, p. 1203).

IC, together with an LF351 connected as an active filter. Figure 5–35 shows the waveforms for the improved circuit.

As discussed, FVCs suffer from a tradeoff of ripple versus speed or settling time. For example, the basic FVC of Fig. 5–33 has about 13 mVp-p of ripple, and a settling time of about 0.6 s, when C_{FILTER} is 1 μF. If you want to increase settling time, decrease the value of C_{FILTER}. However, the ripple will increase by a corresponding amount.

In the circuit of Fig. 5–34, the S/H circuit samples the FVC output at the peak of the ripple, and holds the output until the next cycle. The LF398 has a low output ripple, but does have some short-duration noise spikes and glitches that can be removed with an output filter. The ripple at the output of the active filter is typically less than 1 mV peak. Also, the settling time for a step change of input frequency is about 60 ms, or ten times quicker than the basic FVC of Fig. 5–33 (when C_{FILTER} is 1 μF).

5.3.1 Operation of Improved Circuit

As shown in Fig. 5–35, when the input-frequency waveform has a negative-going transition, pin 6 of the LM331 is driven momentarily lower than the 13-V threshold voltage at pin 7. This initiates a timing cycle controlled by the R_t and C_t at pin 5, and causes a transition from +5 V to 0 V at pin 3 (the normal VFC logic output), which is usually left unused in FVC operation.

During the timing cycle ($t = 1.1 \times R_t \times C_t = 75$ ms, for the example shown) a precision current source ($I = 1.9$ V/R_S) flows out of pin 1 of the LM331, and charges V1 up to a value slightly higher than the average DC value of V1. At the end of the timing cycle, V1 stops charging, and V2 rises. The 10-k pullup resistor is coupled (through the 200-pF capacitor) to V3, and causes the LF398 to sample for about 5 ms.

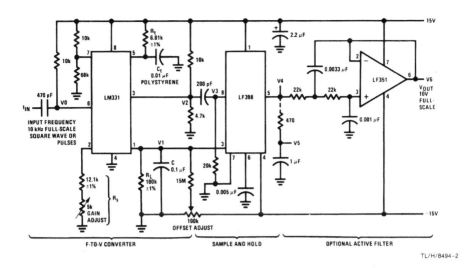

TL/H/8494-2

Figure 5–34. FVC with sample-and-hold (National Semiconductor, *Linear Applications Handbook,* 1994, p. 1203)

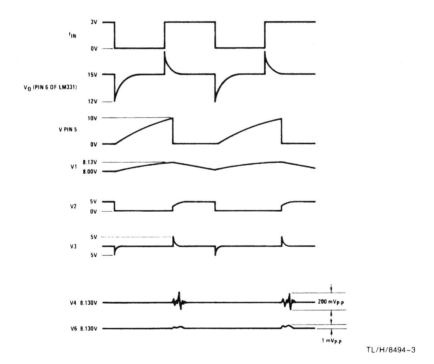

TL/H/8494-3

Figure 5–35. Waveforms for sample-and-hold FVC (National Semiconductor, *Linear Applications Handbook,* 1994, p. 1205)

Then the LF398 goes back to hold. This entire operation is repeated at the same frequency as F_{IN}. The average voltage at V1 will be the same 10-V full-scale (as calculated using the equation of Fig. 5–33). Also, the peak-to-peak ripple (about 130 mVp-p) can be computed using the equation of Fig. 5–33.

The V1 input to the S/H at pin 3 might have a 10.000-V average value, but the output from the S/H will be at 10.065 V because the sample occurs at the peak value of V1. Thus, to get an output with a low offset, a 15-M resistor is used to offset the V1 signal to a lower level.

To calibrate the circuit of Fig. 5–34, start by trimming the offset-adjust pot for a 1-V output with 1 kHz at F_{IN}. Then apply 10 kHz at F_{IN} and trim the gain-adjust pot for a 10-V output. There might be some minor interaction between these two adjustments, as measured at V4, V5, or V6.

In some cases, the simple RC filter (shown at V4) is sufficient to provide a good output at V5. The rms value of the ripple at V4 is small, but the peak-to-peak ripple (including glitches and spikes) can be excessive. If ripple (or noise) is a problem, use the simple active filter (the LF351) to provide a sub-microvolt peak ripple output at V6. The active filter will also improve settling time and provide lower output impedance. Figure 5–35 shows a comparison of the V4 (before filtering) and V6 (after filtering outputs).

Note that the FVC of Fig. 5–34 has good linearity (better than 0.1%), but only from 10 kHz down to 500 Hz. Between 200 Hz and 20 Hz, the output is not truly proportional to F_{IN}. At 0 Hz, the output is indeterminate, because the sample-and-hold will not sample. However, there are many FVC applications where a 20:1 frequency range (500 Hz to 10 kHz) is adequate.

CHAPTER 6

Miscellaneous VFCs and FVCs

This chapter is devoted to simplified-design approaches for various voltage-frequency conversion circuits, as well as for current-to-frequency conversion applications. All of the general design information in Chapter 1 applies to the examples in this chapter. However, each voltage-frequency IC has special design requirements. The circuits in this chapter can be used immediately the way they are or, by altering component values, as a basis for simplified design of similar voltage-frequency conversion applications.

6.1 Current-to-Frequency Converters

Current-to-frequency converters (CFCs) can be formed using VFCs, both IC and discrete component. This is because a VFC actually operates with an input current that is proportional to the voltage input, where $I_{IN} = V_{IN}/R_{IN}$. For example, as shown in Fig. 6–1, I_{IN} is integrated by an op amp, and a charge dispenser (capacitor) acts as a feedback path, to balance out the average input current.

When an amount of charge $Q = I \times T$ (or $Q = C \times V$) per cycle is dispensed by the circuit of Fig. 6–1, the frequency will be:

$$f = (V_{IN} - VOS/R_{IN} + I_b) \times 1/Q$$

When V_{IN} is large in relation to the other factors shown in Fig. 6–1, the frequency equation can be simplified to:

$$f = V_{IN}/R_{IN} \times I/Q$$

When V_{IN} covers a wide dynamic range, the V_{OS} and I_b of the op amp must be considered, because both factors greatly affect the usable accuracy when the input signal is very small. For example, when the full-scale input is 10 V, a signal that is 100 dB below full scale will be only 100 μV. If the op amp has an offset drift of ±100 μV (whether caused by time or temperature), this would cause a ±100% error at this

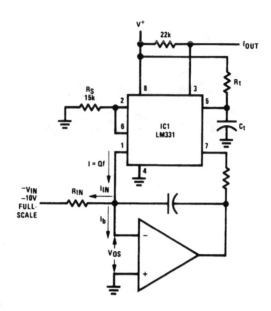

Figure 6-1.
Typical VFC (National
Semiconductor, *Linear
Applications Handbook*,
1994, p. 489)

TL/H/5622-1

signal level. However, a current-to-frequency converter can easily cover a 100 dB range because the voltage offset problem is not significant when the input signal is actually a current source.

6.1.1 Practical Wide-Range CFC

Figure 6–2 shows a practical wide-range CFC using an LF351 op amp. This op amp has an I_b of less than 100 pA. The temperature coefficient of I_b is less than 10 pA/°C at room temperature. The leakage to the charge-dispenser capacitor C_F from the current-source output of the LM331 (at pin 1) is typically 2 pA to 4 pA, and is always less than the 100 pA of the LF351 (at 25°C).

Capacitor C_F must be of a low-leakage type, such as a polypropylene or polystyrene. (At any temperature above about 35°C, the leakage of a mylar capacitor can be excessive.) Also, low-leakage diodes are recommended to protect the circuit input from any possible fault condition. (A 1N914 diode might leak 100 pA, even with only 1 mV, so 1N914s must be avoided.)

After trimming the circuit for a low offset when I_{IN} is 1 nA, the circuit will operate with an input range of 120 dB from 200 mA to 100 pA. This is with an accuracy or linearity error well below 0.02% of the signal, plus 0.0001% of full scale.

The zero-offset drift will be below 5 or 10 pA/°C. As a result, when the input is 100 dB down from full scale, the zero drift will be less than 2% of signal, for a ±5°C temperature range. Another way of indicating this performance is to realize that when the input is 1/1,000 of full scale, zero drift will be less than 1% of that small signal for a 0°C to 70°C temperature range.

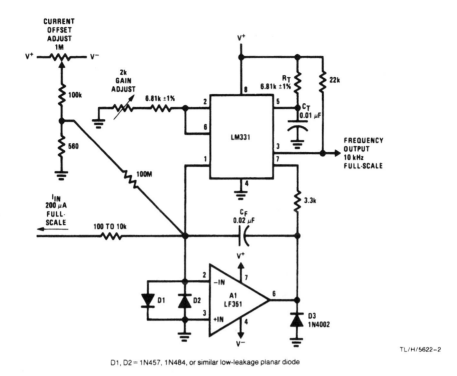

D1, D2 = 1N457, 1N484, or similar low-leakage planar diode

Figure 6–2. Practical wide-range CFC (National Semiconductor, *Linear Applications Handbook,* 1994, p. 490)

6.1.2 Very-Wide-Range CFC

Figure 6–3 shows a very-wide-range CFC using an LF351 op amp, with an FET input circuit to reduce input current. (This configuration is the same as using an expensive op amp with low input current.) The 2N5909 FETs have a maximum I_b of 1.0 pA, and a 1.0 pA/°C drift at room temperature. Typical drift is 0.02 pA/°C.

The voltage-offset adjust pot is used to bring the summing point within a millivolt of ground. With an input signal large enough to cause an output frequency of 1 Hz, trim the voltage-offset pot so that closing the test switch has no effect on the output frequency (or the output period). Then adjust the input-current offset pot to get an output frequency equal to 1/1,000 of full scale, when the input current I_{IN} is 1/1,000 of full scale.

When I_{IN} covers the 140-dB range, from 200 mA to 20 pA, the output will be stable, with very good zero-offset stability, for a limited temperature range near room temperature. However, the following precautions should be observed.

Operate the LM331 on 5-V or 6-V supplies. This will keep leakage down and cut the dissipation (along with the usual temperature rise) to a minimum.

Operate the FETs with a 6-V drain supply.

Guard all summing-point wiring away from all other voltages.

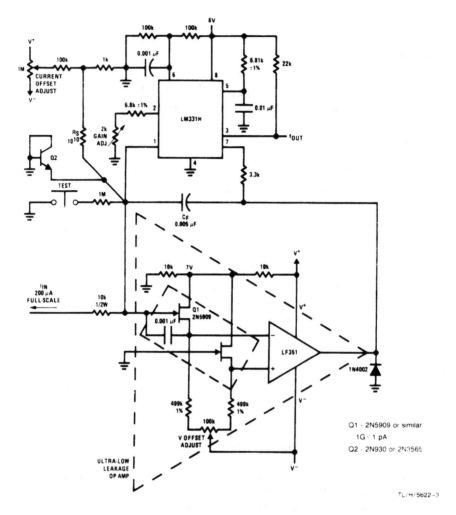

Figure 6–3. Very-wide-range CFC (National Semiconductor, *Linear Applications Handbook,* 1994, p. 491)

6.1.3 Very-Wide-Range CFC with Low Voltage Drift

Figure 6–4 shows a very-wide-range CFC using an LM11C at the input to reduce voltage drift. Typically, voltage drift for the LM11C is less than 2 mV/°C. Current drift is less than 1 pA/°C by itself, and 0.2 pA/°C when trimmed with the 2N3904 bias-compensation circuit shown.

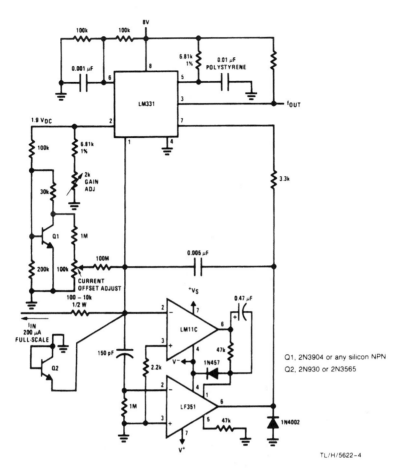

Figure 6–4. Very-wide-range CFC with low voltage drift (National Semiconductor, *Linear Applications Handbook,* 1994, p. 492)

6.1.4 Picoampere-to-Frequency Converter

Figure 6–5 shows a CFC where the full-scale input is 1 mA, for a full-scale output frequency of 100 kHz. This response to currents in the picoampere range is made possible by a current attenuator circuit. The LM334 (a temperature-to-current IC) causes 120-mV bias to appear at the base of Q2. When a current flows out of pin 1 of the LM331, 1/100 of the current will flow out of the Q1 collector, and the remainder of the current flows from the Q2 collector. Because the LM334 current is linearly proportional to Kelvin temperature, the –120 mV at the base of Q2 changes linearly with temperature. As a result, the 1/100 ratio of the Q1/Q2 current divider remains the same across the temperature range.

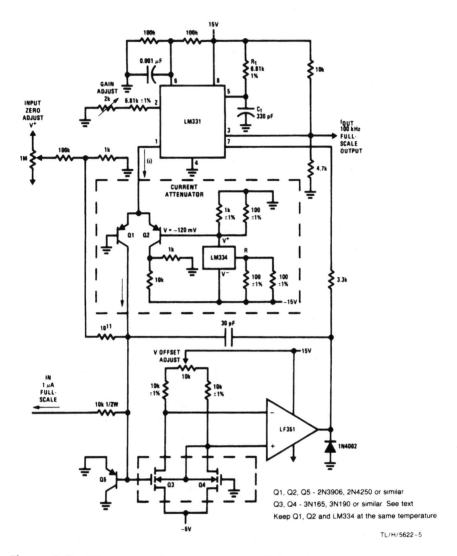

Figure 6–5. Picoampere-to-frequency converter (National Semiconductor, *Linear Applications Handbook,* 1994, p. 493)

The current-attenuator circuit works accurately at high speeds (such as for 4-ms) current pulses. Also, the LM331 leakage is attenuated by a factor of 100 or more using the attenuator. The Q1 leakage is typically 0.01 pA, if the summing point is within a microvolt or two of ground. As indicated by the notes on Fig. 6–5, 3N165 or 3N190 MOSFETs should be used for Q3 and Q4. These MOSFETs have no gate-protection diodes, but do have superior current-leakage and current-drift characteristics.

For adjustment purposes, the full-scale output frequency will be 100 kHz when the input current is 1 mA. At a 1-nA input, the output frequency will be 100 Hz. When the input current is 1 pA, the output frequency drops to 1 cycle per 10 seconds or 100 mHz. The usable dynamic range is better than 140 dB, with accuracy at inputs between 100% and 1%, and between 0.01% and 0.0001%, of full-scale.

6.1.5 CFC for Positive Signals

Figure 6–6 shows a CFC for positive input signals (positive currents) using a current reflector circuit. This 3-transistor reflector provides high output impedance and low leakage. The output can go directly to the summing point, or through a current attenuator using NPN transistors (instead of the PNP transistors used in Fig. 6–5).

The overall circuit output is about 130 dB, with an accuracy of 0.1%. The advantage of wide dynamic range is increased resolution. For example, a 100-kHz output full-scale frequency, instead of 10 kHz, means that resolution is increased by 10

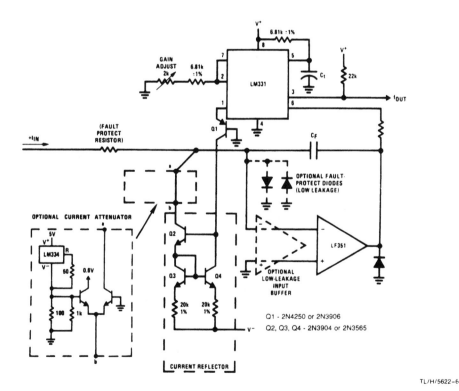

TL/H/5622–6

Figure 6–6. CFC for positive signals (National Semiconductor, *Linear Applications Handbook,* 1994, p. 494)

times. If F_{IN} is 10 Hz with a 100-kHz full-scale, and f_{IN} is integrated or counted for 10 seconds, you can resolve the signal to within 1%.

6.2 Additional FVC Applications

This section is devoted to simplified design of various FVC circuits using the basic VFC ICs discussed in Chapters 1 through 5. Figure 6–7 shows an LM331 IC (or LM131 for the military temperature range) in a basic FVC configuration (the same as shown in Fig. 1–2).

The circuit of Fig. 6–7 is sometimes called a stand-alone converter because no additional active devices (such as op amps) are required. Comparable VFC ICs, such as RM4151 and XR-4151, can take advantage of this and other circuits described here. (However, other VFC ICs might not be pin compatible.)

The circuit of Fig. 6–7 accepts a pulse-train or square-wave input amplitude of 3 V or greater. The 470-pF coupling capacitor suits negative-going input pulses between 1.5 ms and 80 ms, and accommodates square waves or positive-going pulses. However, the time interval between pulses must be at least 10 ms.

6.2.1 Basic LM331 Circuit Operation

The LM331 detects an input-signal change by sensing when pin 6 goes negative, relative to the threshold voltage at pin 7, which is nominally biased 2 V lower

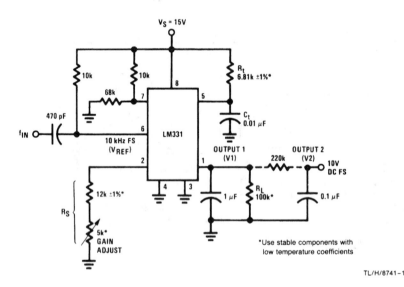

TL/H/8741–1

Figure 6–7. Stand-alone FVC (National Semiconductor, *Linear Applications Handbook,* 1994, p. 1241)

than the supply voltage. When a signal change occurs, the LM331 input comparator sets an internal latch and initiates a timing cycle. During this cycle, a current equal to V_{REF}/R_S flows out of pin 1 for a time $f = 1.1R1C$. The 1-μF capacitor filters this pulsating current from pin 1, and the current's average value flows through load resistor R_L. As a result, for a 10-kHz input, the circuit outputs 10 V across R_L with a typical 0.06% nonlinearity.

6.2.2 Ripple and Response Time

As discussed throughout this book, the basic FVC circuit has two related problems that must be approached with a compromise in mind. The basic circuit of Fig. 6–7 has about 13 mVp-p of ripple, and lags 0.1 second behind an input-frequency step change, settling to within 0.1% of full scale in about 0.6 second. Both of these conditions can be corrected (but with a tradeoff) as follows.

Increasing the filter-capacitor value reduces ripple but also increases response time. Conversely, lowering the filter-capacitor value improves response time at the expense of larger ripple. In most cases, the problem is solved by adding an active filter. This results in faster response time, and less ripple, for high input frequencies.

In cases where ripple is critical, but response time is not, the addition of a passive filter can produce satisfactory results. For example, adding a 220-k/0.1-μF post-filter (as shown in the dotted lines) slows the response slightly, but also reduces ripple to less than 1 mVp-p for frequencies from 200 Hz to 10 kHz.

6.2.3 Power Requirements and Output Load

Although the circuit specifies a 15-V power supply, any regulated supply between 4 V and 40 V can be used. The output voltage can extend to within 3 V of the supply voltage, so choose output-load resistor RL to maintain that output range.

6.2.4 Improving Linearity

Figure 6–8 shows an improvement to the basic FVC circuit. The addition of transistor Q1 increases linearity (or improves nonlinearity) from a typical 0.06% to a typical 0.006%. This 10-to-1 improvement is made possible because Q1 acts as a cascade so that the output impedance at pin 1 sees a constant voltage. With the basic circuit (Fig. 6–7), the current source that feeds pin 1 is turned off most of the time when the input frequency is low. As the input frequency increases, the current source stays on more of the time. This attenuates the output signal for an increasing fraction of each cycle time. The uneven attenuation at higher input frequencies causes a corresponding uneven output (nonlinearity).

6.2.5 Output Buffer and Active Filter

Figure 6–9 shows an output buffer added to the basic circuit. This buffer also acts as an active filter. Either an LM324 or LM358 op amp functions well in the single-supply circuit. This is because the common-mode ranges of such IC op amps ex-

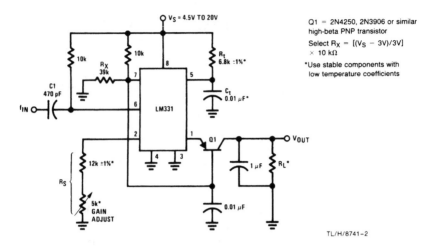

Figure 6–8. Improvement to basic FVC (National Semiconductor, *Linear Applications Handbook,* 1994, p. 1242)

tend down to ground. However, if a negative supply is also available, any op amp can be used. Op amps such as the LF351B or LM308A, which have low input currents, provide the best accuracy.

The two-pole response of the active filter can be expressed as:

$$V_{OUT}/I_{OUT} = R_L/(1 = K1p + K2p^2)$$

where p = the differential operator d/dt. As a result, the DC gain of the filter is controlled by RL. The high-frequency response rolls off at 12 dB/octave. Near the cir-

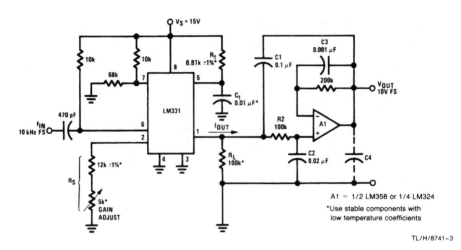

Figure 6–9. FVC with output buffer/filter (National Semiconductor, *Linear Applications Handbook,* 1994, p. 1242)

cuit's natural resonant frequency, the damping can be selected to provide some over-shoot if desired.

6.2.6 Dealing with Ripple

As discussed, adding active filters to the output of an FVC is a standard method of reducing ripple. Figure 6–10 shows a comparison of ripple outputs for the various circuits described in this section. (Note that Figure 1 referred to in Fig. 6–10 is Fig. 6–7 in this book. Figure 3 in Fig. 6–10 is Fig. 6–9 in this book.) Figures 6–11 and 6–12 show inverting and non-inverting filters, respectively, for an FVC output. Both are two-pole filters, and can be cascaded to the FVC output.

The inverting filter of Fig. 6–11 requires closely matched resistors with a low TC over their temperature range for best accuracy. For lowest DC error, choose R5 = R2 + (R$_{IN}$/R$_F$). The response for the inverting filter is:

$$-V_{OUT}/V_{IN} = n/(1 + (R_F + R2 + nR2) \, C4p + R_FR2C3C4p^2)$$

where n = DC gain. If you make R$_{IN}$ = R$_F$, and n = 1, then

$$-V_{OUT}/V_{IN} = 1/(1 + (R_F + 2R2) \, C4p + R_FR2C3C4p^2).$$

The non-inverting filter of Fig. 6–12 does not require precision passive compo-nents. However, for best accuracy, choosing an A1 with a high CMRR is critical. An LM308A op amp (with a 96 dB minimum CMRR) is well suited for this circuit. An LM358B (85 dB CMRR) is also suitable for most applications. The circuit response is:

$$V_{OUT}/V_{IN} = 1/(1 + (R1 + R2) \, C2p + R1R2C1C2p^2)$$

For best results, choose R3 = R1 + R2.

As shown in Fig. 6–10, the simple, slow RC filter (Fig. 6–7) shows reasonably low ripple at all frequencies. The two-pole filters offer the lowest ripple at high fre-quencies, and provide a 30-times-faster step response than the RC configurations.

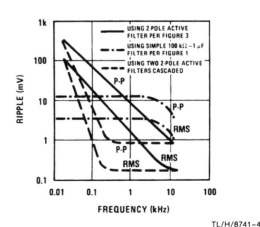

Figure 6–10.
Comparison of ripple out-puts for various circuits (National Semiconductor, *Linear Applications Handbook,* 1994, p. 1243)

TL/H/8741–4

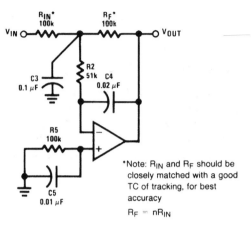

Figure 6–11.
Inverting filter for a FVC output (National Semiconductor, *Linear Applications Handbook,* 1994, p. 1243)

*Note: R_{IN} and R_F should be closely matched with a good TC of tracking, for best accuracy

$R_F = nR_{IN}$

TL/H/8741–5

To reduce ripple at moderate frequencies, cascade a second active-filter circuit (Fig. 6–11 or 6–12) at the output. As shown in Fig. 6–10, the second active-filter approach provides considerable improvement over the single active-filter technique (Fig. 6–9), with only a 30% degradation of the step response.

6.2.7 Simplified Component Selection

The specific response of the circuit in Fig. 6–9 is:

$$V_{OUT}/I_{OUT} = R_L/(1 + (RL + R2)\, C2p + R_L R2C1C2C3C2p^2)$$

However, this rather tedious computation can be simplified by observing the following rules.

The response curves shown in Fig. 6–10 are based on the values shown in Figs. 6–7, 6–9, 6–11, and 6–12.

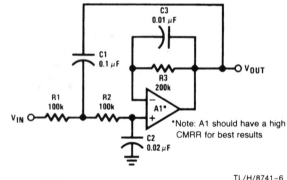

Figure 6–12.
Non-inverting filter for a FVC output (National Semiconductor, *Linear Applications Handbook,* 1994, p. 1243)

*Note: A1 should have a high CMRR for best results

TL/H/8741–6

By maintaining the C1:C2 and R2:RL ratios shown in Fig. 6–9, you can adjust the single two-pole filter to a wide frequency range without computation. Making C2 relatively large eliminates overshoot and sine peaking. Alternatively, making C2 a suitable fraction of C1 (0.02 for C2 and 0.1 for C1) produces both a sine response with 0 dB to 1 dB of peaking, and a quick real-time response, having only 10 to 30% overshoot for a step response.

The filter of Fig. 6–9 settles to within 1% of a 5-V step's final value in about 20 ms. By contrast, the simple RC filter of Fig. 6–7 takes about 900 ms to get the same response, but still has the same ripple as the op-amp circuit of Fig. 6–9.

Any capacitance between 100 pF and 0.05 μF is suitable for C3 in the Fig. 6–9 circuit. This is because C3 serves only as a bypass for the 200-k resistor. Capacitor C4 helps reduce output ripple (in circuits with a single positive supply) when V_{OUT} approaches so close to ground that the op-amp output impedance suffers. In the circuit of Fig. 6–9, using a tantalum capacitor (with a value between 0.1 μF and 2.2 μF) for C4 usually helps keep the filter output much quieter without degrading the op amp stability.

If the circuit of Fig. 6–9 need not go below 100 Hz, make C1 = 10 μF and C2 = 2 μF. Use a tantalum or aluminum electrolytic for C1. However, C2 must be a low-leakage type.

6.2.8 Dealing with Temperature Coefficients

FVCs often have temperature-related problems that are the result of passive-component temperature coefficients (TCs). Use the following guidelines to keep such problems to a minimum.

Teflon or polystyrene capacitors usually show a TC of –110 ±30 ppm/°C. If such capacitors are used as timing capacitors (such as C_t in Fig. 6–13), the output voltage (or gain in terms of volts per kHz) also shows a corresponding TC. The effect of the C_t temperature coefficient is offset by the addition of a resistor-diode network at pin 2. When $R_X = 240$ k, the current flowing through pin 1 will have an overall TC of 110 ppm/°C, effectively canceling the TC of the timing capacitor. This eliminates the need for a zero-TC timing capacitor. However, C_t should have a stable TC. The resistor-diode network also compensates (almost) for the TC of the remaining circuit components.

After the circuit has been assembled, and checked out at room temperature, a brief oven test will indicate the sign and the size of the TC for the complete FVC. You can then add resistance in series with R_X, or add conductance in parallel with RC, to reduce the TC to a minimum.

For example, if the circuit increases the full-scale output by 0.1% per 30°C (33 ppm/°C) during the oven test, adding 120 k in series with $R_X = 240$ k cancels the temperature-caused deviation. Or, if the full-scale output decreases by –0.04% per 20°C (–20 ppm/°C), add 1.2 M in parallel with R_X.

To allow trimming in both directions, start with a finite fixed TC (such as the –110 ppm/°C of Ct). Then cancel the fixed TC with an adjustable TC. This procedure

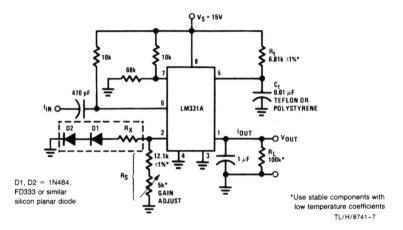

Figure 6-13. FVC with TC-control circuit (National Semiconductor, *Linear Applications Handbook*, 1994, p. 1244)

makes it possible to compensate for whatever polarity of TC is found by the oven test, and to get TCs as low as 20 ppm/°C, possibly even 10 ppm/°C. Consider the following guidelines for best results in TC trimming.

Use a good capacitor for C_t. The cheapest polystyrene capacitors can shift value by 0.05% or more per temperature cycle. Using such a capacitor makes it impossible to distinguish the actual temperature sensitivity from hysteresis.

After soldering, bake or temperature-cycle the circuit (at a temperature not exceeding 75°C in the case of polystyrene) for a few hours to stabilize all components and to relieve the strains of soldering.

Do not rush the trimming. Recheck the room temperature value before and after the high-temperature information is taken. This procedure will ensure a reasonably low hysteresis per cycle.

Do not expect a perfect TC at –25°C if you trim for a ±5 ppm/°C at temperatures from +25°C to 60°C. None of the components in the circuit of Fig. 6–13 offer linearity much better than 5 ppm/°C or 10 ppm/°C cold, if trimmed for a zero TC at warm temperatures. Even so, it is still possible to get a data-converter circuit with an 0.02% accuracy and 0.003% linearity, for a ±20°C range near room temperature.

Start the trimming with R_X installed, and at a value near the design-center value (240 k or 270 k). Such values should produce a TC near zero. Do not start trimming without an R_X installed.

If you change R_X (say, from 240 k to 220 k), do not pull out the 240-k resistor, and put in a new 220-k part. The results will be more consistent if a 2.4-M resistor is added in parallel. This admonition is also true when adding resistances in series with R_X.

Use reasonably stable components. (This is a good idea for simplified design of any data-converter circuit!) If you use an LM331A (±50 ppm/°C maximum) and

RN55D film resistors (each ±100 ppm/°C) for R_L, R_t, and R_S, it will generally be impossible to trim out the worst-case ±350 ppm/°C TC. Resistors with a specification of 25 ppm/°C usually work well.

Use the same resistor for both R_S and R_t. When these resistors come from the same manufacturer's batch, their TC tracking will usually be better than 20 ppm/°C.

Whenever an op amp is used as a buffer (such as in Fig. 6–9), the offset voltage and current (±7.5 mV maximum and ±100 nA, respectively, for most inexpensive IC op amps) can cause a ±17.5-mV worst-case output offset. However, if both plus and minus supplies are available, a symmetrical offset adjustment can be added. With only one supply, a small positive current can be added to each op-amp input, and one input can be trimmed.

6.2.9 FVC with Negative Output

Figure 6–14 shows an FVC with negative output. The linearity is ±0.003% typical, and ±0.01% maximum. Because pin 1 of the LM331 always remains at 0 V, the circuit needs no cascade transistor (as in Fig. 6–8). Although linearity is good, the circuit does show some ripple.

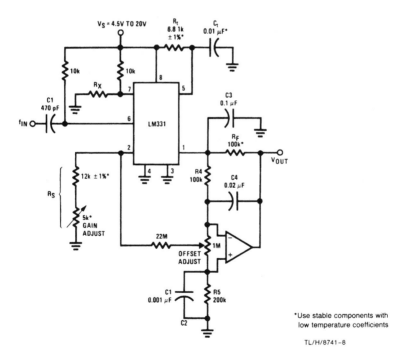

TL/H/8741–8

Figure 6–14. FVC with negative output (National Semiconductor, *Linear Applications Handbook,* 1994, p. 1245)

The main advantage of the Fig. 6–14 circuit is that the offset-adjust voltage is taken from a stable 1.9-V reference at pin 2 of the LM331. Any supply-voltage shifts cause no shift in output. The offset pot can be of any value between 200 k and 2 M.

The optional bypass capacitor around resistor R5 prevents output noise arising from stray noise at the op-amp input. (The capacitance value is not critical.) The best bias-current compensation is obtained when $R5 = R4 + R_F = 200$ k.

The circuit of Fig. 6–14 shows the same 2-pole response, with heavy output-ripple attenuation, as the non-inverting filter in Fig. 6–9. Specifically, the Fig. 6–14 response is:

$$V_{OUT}/I_{OUT} = R_F/(1 + (R4 + R_F) C4p + R4R_FC3C4p^2).$$

6.2.10 FVC with Temperature Sensor

Figure 6–15 shows an FVC with a temperature sensor (LM334) to provide temperature compensation. The circuit can handle input frequencies up to 100 kHz using the capacitor values shown. This circuit increases the current at pin 2 to facilitate high-speed switching. However, the 500 ppm/°C TC of the LM331 (at 100 kHz) can cause problems because of the switching-speed shifts resulting from temperature changes.

To compensate for the positive TC of the LM331, the LM334 sensor feeds pin 2 with a current that decreases linearly with temperature. This decrease in current offsets the positive increase, and provides a low overall TC. An R_y value of 30 k (at pin 2 of the LM331) is generally sufficient for first-order compensation. However, other

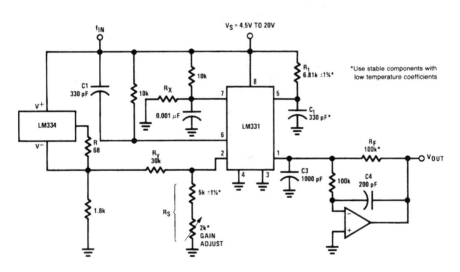

Figure 6–15. FVC with temperature sensor (National Semiconductor, *Linear Applications Handbook,* 1994, p. 1246)

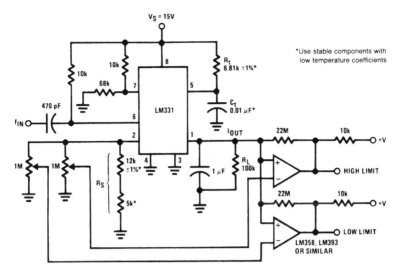

Figure 6–16. FVC combined with comparators to form a frequency detector (National Semiconductor, *Linear Applications Handbook,* 1994, p. 1246)

values of R_y can be used for more precise trimming (with oven testing as described in Sec. 6.2.8).

6.2.11 Slow-Response Frequency Detector

Figure 6–16 shows an FVC combined with two comparators to form a frequency detector. The comparators are adjusted by the 1-M pots so that they switch states when the input signal goes above or below a given frequency. Although the circuit is accurate, the response is slow.

With the values shown, a frequency drop from 1.1 kHz to 0.5 kHz will cause a change in states at the comparator output in about 20 ms. When the input frequency falls from 9 kHz (near full-scale) to 0.9 kHz, the output responds only after a 600-ms lag. As a result, the circuit can be used only where inherent delays (and ripple) can be tolerated.

6.3 Miscellaneous Voltage-Frequency Converter Applications

This section is devoted to simplified design of various circuits using the basic VFC IC (the LM331). The circuits include oscillators, telemetry converters, and analog computers, as well as an ultralinear VFC.

6.3.1 Voltage-Controlled Relaxation Oscillator

Figure 6–17 shows the basic VFC connected as a voltage-controlled relaxation oscillator. (Note that this circuit is the same as shown in Fig. 1–1.) The LM331 is basically a precision relaxation oscillator that generates a frequency linearly proportional to the input voltage. (This is true of virtually all IC VFCs.)

In effect, the circuit is a feedback loop that keeps capacitor C_L charged to a voltage slightly higher than the input voltage (V_{IN}). The voltage across C_L has a sawtooth waveform. If V_{IN} is high, C_L discharges quickly through R_L, and the circuit generates a high frequency. If V_{IN} is low, C_L discharges slowly, and the output is at a low frequency.

When C_L discharges to a voltage equal to the input, the comparator triggers the one-shot. As a result, the one-shot closes the current switch, and turns on the output transistor. With the switch closed, current from the current source recharges C_L to a voltage somewhat higher than V_{IN}. Charging continues for a period determined by R_T and C_T. At the end of this period, the one-shot returns to the quiescent state and C_L resumes discharging.

Resistor R_S sets the amount of current put out by the current source. The current in pin 1, with the switch on, is identical to the current in pin 2. The nominal voltage on pin 2 is 1.9 V, so a given resistor value sets the operating currents. When connected to a high-impedance buffer, pin 2 provides a stable reference for external circuits.

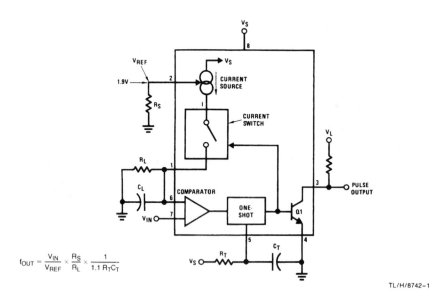

$$f_{OUT} = \frac{V_{IN}}{V_{REF}} \times \frac{R_S}{R_L} \times \frac{1}{1.1\, R_T C_T}$$

TL/H/8742–1

Figure 6–17. Voltage-controlled relaxation oscillator (National Semiconductor, *Linear Applications Handbook*, 1994, p. 1247)

The open-collector output at pin 3 permits the output swing to be different from the supply voltage (V_S) if the load circuit requires different values. However, both the converter and the load can use the same voltage.

6.3.2 Stable-Frequency Oscillator

Figure 6–18 shows the basic VFC connected as a stable-frequency oscillator. This circuit ties the reference output voltage back to the signal input (pins 2 and 7). The circuit then acts as a VFC with a constant input, and results in a constant-frequency oscillator output.

In the circuit of Fig. 6–18, variations in reference voltage have two opposite effects that cancel each other out, thus producing frequency and temperature stability. The temperature-dependent internal delays also tend to cancel. (This is not true of relaxation oscillators based on op amps or comparators.)

Resistors R_L and R_S are best taken from the same batch. R_L must be larger than R_S, so R_L is made up with two resistors in series. By taking resistors from the same batch, and using two resistors for R_L, the TC tracking (which is the critical factor) is five to ten times better than would be the case if R_L was a single 30.1-k resistor. Although the reference output at pin 2 cannot be loaded without affecting the converter sensitivity, the comparator input at pin 7 has a high impedance, so this feedback connection does no harm.

Frequency stability for the circuit is typically ±25 ppm/°C, even with an LM331 (which is specified only to 150 ppm/°C when used as a VFC). From 20 Hz to 20 kHz, stability is excellent, and the circuit can generate frequencies up to 120 kHz. Use the equations shown in Fig. 6–18 to find component values for the desired operating or output frequency.

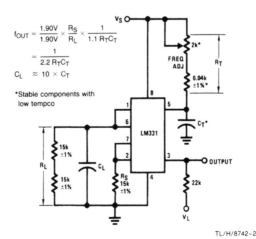

$$f_{OUT} = \frac{1.90V}{1.90V} \times \frac{R_S}{R_L} \times \frac{1}{1.1\,R_T C_T}$$

$$= \frac{1}{2.2\,R_T C_T}$$

$$C_L \approx 10 \times C_T$$

*Stable components with low tempco

Figure 6–18.
Stable-frequency oscillator (National Semiconductor, *Linear Applications Handbook*, 1994, p. 1248)

TL/H/8742–2

6.3.3 Strain-to-Frequency Converter

Figure 6–19 shows the basic VFC combined with a strain gauge to form a strain-to-frequency converter such as those used in telemetry applications. In this circuit, the reference output voltage is buffered and amplified to supply the resistive strain gauge (or a pot transducer). Any deviations of the internal reference voltage from the ideal cause both the transducer and converter sensitivities to change equally in opposite directions, so the effects cancel.

In the circuit of Fig. 6–19, op amp A2 buffers and amplifies the constant voltage at pin 2 of the converter to provide the 5-V excitation for the strain gauge. Amplifier A1, connected as an instrumentation amplifier, raises the output of the strain gauge to a usable level, while simultaneously rejecting common-mode pickup. A potentiometer-type transducer can be used in place of the strain gauge. The wiper output takes the place of the A1 output as shown.

6.3.4 Analog Ratio Computer

Figure 6–20 shows the basic VFC connected as an analog ratio computer. The circuit converts the ratio of two voltages to an equivalent frequency without a separate analog divider. (Note that this circuit is the same as shown in Fig. 1–10.) The circuit makes use of the reference output at pin 2 as a current-programming signal input. The extra input enables the LM331 to compute while converting. In effect, the circuit is still a VFC, but with two signal inputs.

The inputs, shown as voltages, are converted to currents by two current pumps (voltage-to-current converters). If currents of the proper ranges are available, the current pumps can be omitted. The left current pump Q1-A1 determines how fast capacitor C_L discharges between output pulses. The other current pump Q2-A2 sets the current in the reference circuit to control the amount of recharge current when the one-shot fires. The comparator trip point is set at a constant voltage by tying the comparator input (pin 7) to the reference (pin 2).

To understand how the circuit operates, consider the effect if the input voltage V1 is tripled. This makes C_L discharge to the comparator trip point three times as fast, so the frequency triples. Now consider a given change, such as doubling the voltage at the other input V2. This doubles the recharge current to C_L during the fixed-width output pulse, and the C_L voltage increases twice as much during recharging. Because the discharge into Q1 is linear (V1 constant), it takes twice as long for CL to discharge. As a result, the frequency is cut in half.

The current pumps in Fig. 6–20 require negative inputs. Figure 6–21 shows current pumps that provide for positive inputs and for offset trimming. Trimming out the offset in the op amp gives the ratio converter better linearity and accuracy.

The trim circuit in Fig. 6–21a requires stable positive and negative supplies for the offset trimmer. The circuit in Fig. 6–21b needs only a stable positive supply. (Unmarked components in Fig. 6–21b are the same as in Fig. 6–21a.)

Note that the full-scale range of the current pumps can be changed by varying the value of the input resistor(s). If either of the pump circuits is used with a single

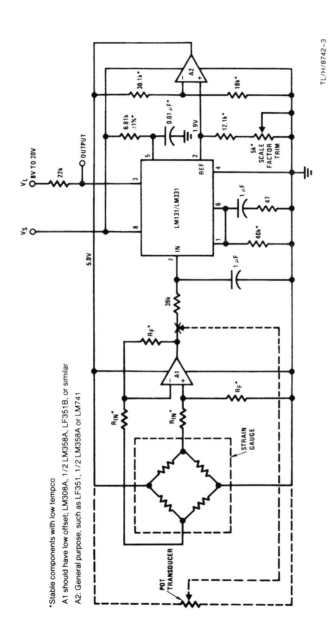

*Stable components with low tempco

A1 should have low offset; LM308A, 1/2 LM358A, LF351B, or similar

A2: General purpose, such as LF351, 1/2 LM358A or LM741

Figure 6–19.
Strain-to-frequency converter (National Semiconductor, *Linear Applications Handbook*, 1994, p. 1248)

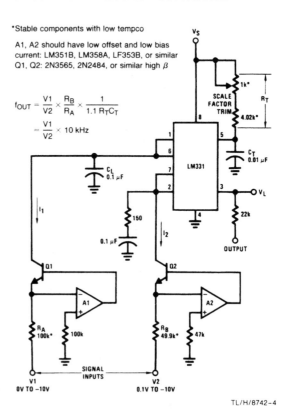

Figure 6–20.
Analog ratio computer
(National Semiconductor,
*Linear Applications
Handbook,* 1994, p. 1249)

TL/H/8742-4

positive supply, the op amp should be a type such as 1/2 LM358 or 1/4 LM324, which have a common-mode range that includes the negative-supply bus.

6.3.5 Analog Square-Root Computer

Figure 6–22 shows a theoretical analog-divider circuit used to compute the square root of an input signal. Such a circuit functions by feeding the input signal to the numerator input of the divider, and returning the divider output back to the divider denominator input. This type of computation is called implicit because the end result of the computation is only implied, not explicitly stated by the equation that defines the computation.

In the implicit square-root computing loop shown in Fig. 6–22, the divider is a VFC, and the output is a frequency. Connecting the output back to one of the inputs so that the circuit will compute a square root means that the output frequency must be converted back to a voltage. This is done by the theoretical circuit of Fig. 6–23.

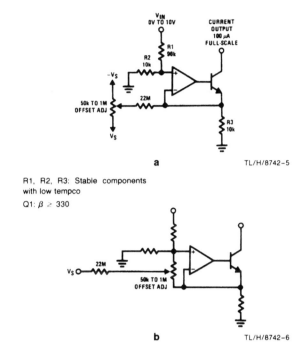

R1, R2, R3: Stable components
with low tempco

Q1: $\beta \geq 330$

Figure 6–21.
Current pumps for positive inputs (National Semiconductor, *Linear Applications Handbook,* 1994, p. 1249)

TL/H/8742-5

TL/H/8742-6

6.3.6 *Analog Multiplying Computer*

Figure 6–24 shows the basic VFC connected as an analog multiplying computer. The circuit multiplies the two inputs while converting to an equivalent frequency. This circuit has a more elaborate current pump than the ratio circuit of Fig. 6–20. The current pump in Fig. 6–24 has two cascaded circuits, and includes op amps A2-A3, as well as transistors Q2-Q3. Current from this pump goes to pin 5 of the VFC to control the pulse width of the one-shot. (This current ranges from 13.3 mA to 1.33 mA.)

As in the ratio circuit of Fig. 6–20, the left current pump of Fig. 6–24 controls the discharge rate of CL. The other pump controls the one-shot pulse width to vary the amount that CL charges during the pulse.

Figure 6–22.
Theoretical analog divider (National Semiconductor, *Linear Applications Handbook,* 1994, p. 1249)

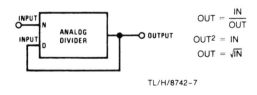

$$OUT = \frac{IN}{OUT}$$

$$OUT^2 = IN$$

$$OUT = \sqrt{IN}$$

TL/H/8742-7

Figure 6–23.
Theoretical analog divider for square roots
(National Semiconductor, *Linear Applications Handbook,* 1994, p. 1250)

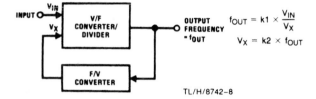

TL/H/8742-8

If the V2 input is near zero, the current from the pump into pin 5 is small, and the one-shot develops a wide pulse. This allows C_L to charge substantially. As a result, it takes a relatively long time for C_L to discharge to the comparator threshold, and the frequency is low. When V2 goes negative (a greater absolute magnitude), the output frequency rises. Op amp A3 must have a common-mode range that extends to the positive-supply voltage. (Use the IC types specified in Fig. 6–24 for best results.)

By combining the V2 input current pump of Fig. 6–20 with the circuit of Fig. 6–24, it is possible to multiply, divide, and convert to a frequency simultaneously. If a scale-factor trimmer is needed, R4 in Fig. 6–24 is a good choice (better than input

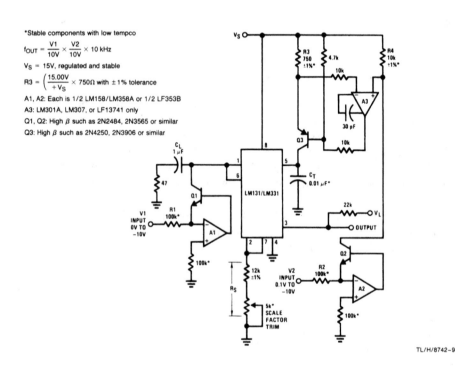

TL/H/8742-9

Figure 6–24. Analog multiplying computer (National Semiconductor, *Linear Applications Handbook,* 1994, p. 1250)

resistors such as R1 or R2). If R1 or R2 is used, the input impedance of the circuit changes with trim setting.

6.3.7 Practical Analog Square-Root Computer

Figure 6–25 shows two VFCs connected to provide an output frequency that is proportional to the square root of the input voltage. (This is a practical version of the theoretical circuit of Fig. 6–23, and includes some of the elements in Fig. 6–24.) ICl is used as a VFC and divider. IC2 is used as an FVC.

IC2 and the current pump that includes A1 return the output of IC1 to the denominator input. IC1 produces a frequency proportional to V_{IN}, divided by the feedback voltage V_X. The current I1 is generated by a current pump that has V_X as the input (Fig. 6–21a). To develop the feedback, IC2 converts the pulse output from IC1 into standard-width precision current pulses that charge capacitor C1. In turn, C1 integrates the current pulses into the voltage V_X, thus closing the loop.

Op amp A2 serves as a comparator to ensure that the circuit will always start and continue running. For example, if V_{IN} suddenly jumps to a higher voltage (a typical step input), one pulse from the one-shot in IC1 might not be enough to recharge C_L to a voltage higher than the input. In such a case, the internal logic of IC1 keeps

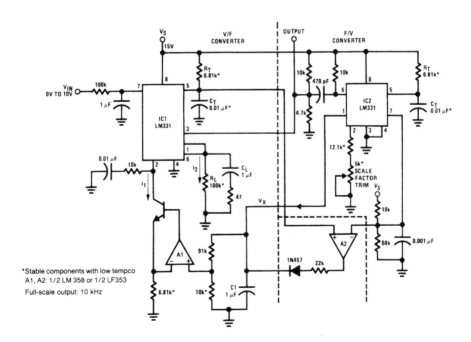

Figure 6–25. Practical analog square-root computer (National Semiconductor, *Linear Applications Handbook,* 1994, p. 1252)

the internal current switch turned on. The voltage on C_L then ramps up until the voltage exceeds V_{IN}.

During the ramp period, the IC1 output does not change state. (This is a typical condition for IC VFCs.) Without a change of state, the lack of pulses to IC2 means that V_X and I_1 decay. The recharging current I_2 for C_L is the same as I_1. As a result, it not only becomes progressively harder for the voltage on C_L to catch up with the input, the C_L voltage might even fail to catch up entirely if $I_2 \times R_L$ is less than the input voltage.

When the circuit of Fig. 6–25 hangs up as just described, the one-shot timing node (pin 5) continues to charge well beyond the normal peak of 2/3 V_S. As soon as the comparator A2 detects this rise, A2 pulls up voltage V_X, current I_1 increases, and the loop continues running (in spite of the initial step input).

6.3.8 Ultraprecision VFC

Figure 6–26 shows the basic VFC connected as a precision voltage-frequency converter. The circuit has many refinements over the other VFCs described in this chapter. Choosing the proper components, and trimming the TC, results in a circuit

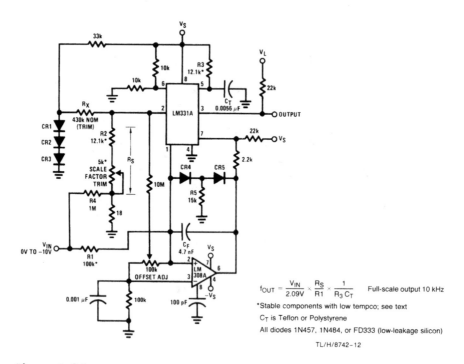

$$f_{OUT} = \frac{V_{IN}}{2.09V} \times \frac{R_S}{R1} \times \frac{1}{R_3 \, C_T}$$ Full-scale output 10 kHz

*Stable components with low tempco; see text

C_T is Teflon or Polystyrene

All diodes 1N457, 1N484, or FD333 (low-leakage silicon)

TL/H/8742–12

Figure 6–26. Precision voltage-frequency converter (National Semiconductor, *Linear Applications Handbook*, 1994, p. 1253)

with less than 0.02% error and 0.003% nonlinearity, for a ±20°C range around room temperature.

The circuit of Fig. 6–26 has an active integrator, which includes the op amp and the integrating feedback-capacitor C_F. The integrator converts the input voltage, which is negative, into a positive-going ramp. When the ramp reaches the comparator threshold of the converter IC, the one-shot fires and switches a pulse of current to the integrator summing junction. This current makes the integrator output ramp down quickly. When the one-shot times out, the cycle repeats.

Here are some reasons that the circuit of Fig. 6–26 provides the high perform-ance. First, a feedback limiter prevents the op amp from driving pin 7 of the VFC negative. The limiter circuit configuration bypasses the leakage through CR5 to ground through R5. As a result, the leakage does not reach the summing junction. By-passing leakage in this way is especially important at high temperatures.

Second, the offset-trimming pot is connected to the stable 1.9-V reference at pin 2, instead of to a power-supply bus that might be unstable and noisy.

Third, a small fraction (180 mV, full-scale) of the input voltage goes to the R_S network through R4. This improves the non-linearity from 0.004% to 0.002%.

Fourth, resistors R2 and R3 are the same value, so that resistors such as Allen-Bradly type CC metal-film types can provide good TC tracking at low cost. (The tracking is best when equal-value resistors come from the same batch.) Resistor R1 should be a low-TC metal-film or wirewound type, with a minimum TC of ±10 ppm/°C or ±25 ppm/°C.

Timing capacitor C_T should be a polystyrene or Teflon type. Polystyrene is rated to 80°C, whereas Teflon goes to 150°C. Both types can be obtained with a TC of –110 ±30 ppm/°C. Choosing this TC for the timing capacitor makes the full-scale output TC (because of C_T) 110 ppm/°C.

Using tight-tolerance components results in a total TC between 0 ppm/°C and 220 ppm/°C, so the TC will never be negative. The voltage at CR1 and R_X has a TC of –6 mV/°C, which can be used to compensate the TC of the remainder of the cir-cuit. Trimming R_X compensates for the TC of the VFC IC, the timing capacitor, and all of the resistors.

A good starting value for selecting R_X is 430 k, which will give the 135 mA flowing out of pin 2 a slope of 110 ppm/°C if the output frequency increases with temperature, add some conductance in parallel with R_X.

When doing a second round of trimming, note that a resistor of 4.3 M has about the same effect on TC (when shunted across a 220-k resistor) as when shunted across a resistor of 430 k (typically –11 ppm/°C). With proper trimming, the TC can be ±20 ppm/°C, or even ±10 ppm/°C.

Here are some recommendations for completing the circuit in final form.

Use a good capacitor for C_T. The cheapest polystyrene capacitors will shift in value by 0.05% or more per temperature cycle. The actual temperature sensitiv-ity would be indistinguishable from the hysteresis, and the circuit would never be stable.

After soldering, bake and/or temperature-cycle the circuit (at a temperature not exceeding 75°C if C_T is polystyrene) for a few hours to stabilize all components and to relieve the strains from soldering.

Do not rush the trimming. Recheck the room-temperature value, before and after the high-temperature information is taken. This will ensure that hysteresis per cycle is reasonably low.

Do not expect a perfect TC at –25°C if the circuit is trimmed for ±5 ppm/°C between 25°C and 60°C. If the circuit is trimmed for zero TC while warm, none of the components will be linear to match better than 5 ppm/°C to 10 ppm/°C when cold.

The values shown in Fig. 6–26 are generally optimum for ±12-V to ±16-V regulated supplied. However, any stable supplies between ±4 V and ±22 V can be used, after changing a few component values.

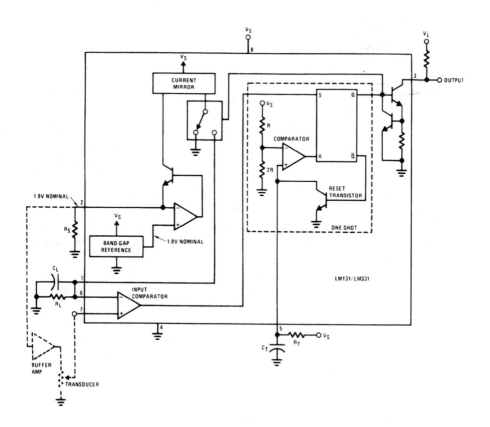

Figure 6–27. Simplified VFC with transducer and external components (National Semiconductor, *Linear Applications Handbook,* 1994, p. 1251)

6.3.9 Design Notes for the LM331/LM131 (a summary)

The following notes should be considered when using the LM331/LM131 IC for any circuit. Figure 6–27 shows a simplified schematic of the basic VFC, with a transducer and some typical external components.

The reference circuit (at pin 2) is both a constant-voltage output and a current-setting, scale-factor control input. The constant voltage can supply external circuits, such as the transducer, that feed the comparator input.

The main advantage of using the internal reference to supply the external circuits is that any variation in the reference voltage affects the sensitivities of the VFC and the external circuits by equal, and opposite, amounts. This cancels the effects of any variation.

In addition to providing a constant-voltage output, pin 2 also provides scale-factor, or sensitivity control, for the VFC. Current supplied to an external circuit from pin 2 comes from the supply V_S through a current mirror and the transistor. The op amp drives this transistor to hold pin 2 at a constant voltage equal to the internal reference (band-gap reference), which is nominally 1.9 V.

The current mirror provides a current (essentially identical to that in pin 2) to the switch. This means that a resistor to ground, or a signal from a current source, will set the current that is switched to pin 1. In most circuits, pin 1 is connected to ground through a capacitor. The current at pin 1 recharges the capacitor during the pulse from the one-shot.

The one-shot circuit is somewhat like the familiar 555 timer circuit. In the quiescent state, the reset transistor is on, and holds pin 5 near ground. When pin 7 becomes more positive than pin 6 (or pin 6 falls below pin 7), the input comparator sets the flip-flop in the one-shot.

The flip-flop turns on the current-limited output transistor (pin 3) and switches the current coming from the current mirror to pin 1. The flip-flop also turns off the reset transistor, and the timing capacitor C_T starts to charge toward V_S. This charge is exponential, and the C_T voltage reaches 2/3 of V_S in about 1.1 $R_T C_T$ time constants. When pin 5 reaches 2/3 of V_S, the one-shot comparator resets the flip-flop, which turns off the current to pin 1, discharges C_T, and turns off the output transistor.

If the voltages at pins 6 and 7 still call for setting the flip-flop after pin 5 has reached 2/3 of V_S, internal logic (not shown) overrides the reset signal from the one-shot comparator. In this case, the flip-flop stays set, and C_T continues charging past 2/3 V_S.

For Further Information

When applicable, the source for each circuit or table is included in the circuit or table title, so that the reader may contact the original source for further information. To this end, the mailing address and telephone and/or fax number for each source is given in this section. When writing or calling, give complete information, including circuit title and description. Notice that all circuit diagrams and tables have been reproduced directly from the original source, without redrawing or resetting, by permission of the original publisher in each case.

AIE Magnetics
701 Murfreeboro Road
Nashville, TN 37210
(616) 244-9024

Analog Devices
One Technology Way
PO Box 9106
Norwood, MA 02062-9106
(617) 329-4700
fax (617) 326-8703

Dallas Semiconductor
4401 S. Beltwood Parkway
Dallas, TX 75244-3292
(214) 450-0400

EXAR Corporation
2222 Qume Drive
PO Box 49007
San Jose, CA 95161-9007
(408) 434-6400
fax (408) 943-8245

GEC Plessey Semiconductors
Cheney Manor
Swindon, Wiltshire
United Kingdom SN2 2QW
0793 518000
fax 0793 518411

Harris Semiconductor
PO Box 883
Melbourne, FL 32902-0883
(407) 724-7000
(800) 442-7747
fax (407) 724-3937

Linear Technology Corporation
1630 McCarthy Boulevard
Milpitas, CA 95035-7487
(408) 432-1900
(800) 637-5545

Magnetics
Division of Spang and Company
900 East Butler
PO Box 391
Butler, PA 16003
(412) 282-8282

Maxim Integrated Products
120 San Gabriel Drive
Sunnyvale, CA 94086
(408) 737-7600
(800) 998-8800
fax (408) 737-7194

Motorola, Inc.
Semiconductor Products Sector
Public Relations Department
5102 N. 56th Street
Phoenix, AZ 85018
(602) 952-3000

National Semiconductor Corporation
2900 Semiconductor Drive
PO Box 58090
Santa Clara, CA 95052-8090
(408) 721-5000
(800) 272-9959

Optical Electronics Inc.
PO Box 11140
Tucson, AZ 85734
(602) 889-8811

Philips Semiconductors
811 E. Arques Avenue
PO Box 3409
Sunnyvale, CA 94088-3409
(408) 991-2000

Raytheon Company
Semiconductor Division
350 Ellis Street
PO Box 7016
Mountain View, CA 94039-7016
(415) 968-9211
(800) 722-7074
fax (415) 966-7742

Semtech Corporation
652 Mitchell Road
Newbury Park, CA 91320
(805) 498-2111

Siliconix Incorporated
2201 Laurelwood Road
Santa Clara, CA 95054
(408) 988-8000

Unitrode Corporation
8 Suburban Park Drive
Billerica, MA 01821
(508) 670-9086

Index